The Mechanical Emotions

Decoding the Robotic Canvas of Facial Expressions

ISHWAR SINGH   BIRINDER PAL KAUR

RAHUL PAWAR   KANCHAN PAWAR

The Mechanical Emotions

By Ishwar Singh and Birinder Pal Kaur

ISBN: 9798867667078 (Paperback)

Cover Design: Ishwar Singh

Interior Design: Ishwar Singh

Publisher: The Enlightenment

Published in United States of America

First Edition: 2023

DEDICATED TO

*I am dedicating this book to my father-in-law and mother-in-law
Sardar Narinder Singh and Late Sardarni Gurjeet Kaur*

FOREWORD

In the ever-evolving field of artificial intelligence, where the borders between technology and mankind blur, "The Mechanical Emotions: Decoding the Robotic Canvas of Facial Expressions" emerges as a riveting excursion into the delicate area of emotional robotics. Authored with careful insight and creative foresight, this book navigates the unexplored seas of humanoid robots and their potential to convey and interpret emotions via facial expressions.

As we stand at the crossroads of innovation and contemplation, this book takes readers on a thought-provoking journey into the core of human-robot interaction. In a society increasingly connected with technology, the research of artificial emotions is not only a scientific pursuit but a deep meditation on what it means to be human in the digital era.

The writers, with their experience and enthusiasm, explore the architecture of robotic faces, uncover the difficulties of emotion replication in machines, and show the route towards programming emotions using algorithms and artificial intelligence. This book serves as a guiding compass through the problems and possibilities posed by the merger of human emotions with robotic capabilities.

Through thorough case studies, the story expands beyond theoretical frameworks, revealing the real-world influence of emotional robots in healthcare, education, and daily life. These examples uncover the transformational potential of emotional robotics, illustrating the practical advantages that may be gained when technology and empathy intersect.

Yet, among the promises of advancement, the writers expertly address the ethical problems that accompany the advent of emotional robots. From privacy issues to the possible loss of employment, the book addresses the challenges head-on, proposing a nuanced and ethical approach to the development and integration of this transformational technology.

"The Mechanical Emotions" is not only a scientific discourse; it is a call to focus on the cultural ramifications of a future where humans and emotional robots cohabit together. The authors ask us to contemplate the cultural subtleties that affect our acceptance of these robots and discuss the ethical imperatives that should drive their progress.

As we dig into the pages of this book, we are pushed to picture a future where emotional robots improve our lives without diminishing the core of our humanity. The authors give us with the means to navigate this future wisely, asking us to embrace the opportunities while conserving the values that define us.

In a world where technology continues to change our interactions with the environment around us, "The Mechanical Emotions" stands as a light, illuminating the path ahead. It is an intellectual voyage that awakens curiosity, promotes thinking, and invites conversation. Whether you are a seasoned specialist in robotics or a curious observer of the technological environment, this book calls you to go on a journey that transcends disciplines and connects with the essence of what it means to be alive in an era of mechanical emotions.

Er. Janardhan Kumar

PREFACE

Welcome to "The Mechanical Emotions: Decoding the Robotic Canvas of Facial Expressions." In the enormous terrain of human-machine interaction, the investigation of robotic emotions stands as a compelling frontier. This book goes on a trip into the complicated domain of face expressions in humanoid robots, where the combination of technology and emotion offers a canvas filled with possibilities.

The obsession with humanoid robots that can exhibit emotions has been a driving factor in both scientific and public debate. From the early days of robotics to the cutting-edge breakthroughs of today, the struggle to read the delicate subtleties of facial emotions in artificial entities echoes our drive to bridge the gap between human and machine. This book tries to investigate the mysteries, problems, and promises inherent in the goal of imbuing robots with the capacity to transmit and comprehend emotions via their synthetic visages.

As we browse the chapters herein, we will dig into the anatomy of robotic faces, investigating the various design ideas and technologies that enable the production of expressive robots. From the algorithms driving emotion replication to the ethical issues in constructing emotionally intelligent robots, each piece unravels a layer of the intricate tapestry that is the junction of robotics and emotions.

Human-robot interaction dynamics, the programming of emotions via artificial intelligence, and the enormous ramifications of emotional robots in healthcare, education, and beyond are all investigated under the lens of this inquiry. Through case studies and real-world applications, we hope to explain how these mechanical emotions materialize in realistic circumstances, influencing the landscape of our shared future with robots.

The notion of a happy future between humans and empathetic robots is a frequent motif throughout this voyage. We examine the ethical imperatives, social repercussions, and the obstacles of acceptance that precede the incorporation of robots into our lives. This book does not only strive to decipher the technical elements of facial expressions in robots but seeks to provoke discussion on the enormous cultural, ethical, and emotional consequences that follow this emerging topic.

As we immerse ourselves in the realm of "The Mechanical Emotions," it is my intention that this investigation acts as both a guide and a spark for greater debates. Whether you are a seasoned researcher, an enthusiast in the area, or an inquisitive thinker wanting to comprehend the changing link between people and machines, this book is created to be a companion in your trip through the complicated world of artificial emotions.

Let the pages ahead unravel the complex tale of how we decipher the robotic canvas of facial expressions, and may this trip excite your curiosity, challenge your preconceptions, and stimulate thought on the ever-evolving tapestry of human-robot connection.

Ishwar Singh

x

ACKNOWLEDGMENTS

I would like to convey my deepest thanks to my father-in-law, Sardar Narinder Singh, and late mother-in-law, Sardarni Gurjeet Kaur, for their steadfast support and encouragement throughout the construction of "The Mechanical Emotions: Decoding the Robotic Canvas of Facial Expressions."

Sardar Narinder Singh's knowledge and leadership have been a source of inspiration along this trip. His intelligent chats and ongoing support have played a vital part in molding the thoughts and concepts offered in this book. His conviction in the significance of studying the interplay of technology and human emotions has been a motivating factor behind my work.

Although Sardarni Gurjeet Kaur is no longer with us, her memory remains a treasured presence. Her warmth, generosity, and love have made an unforgettable impression on my life. I am thankful for the ideals she established and the support she offered, which have shaped the spirit of this work.

To both Sardar Narinder Singh and the late Sardarni Gurjeet Kaur, I send my heartfelt thanks for being pillars of strength, creating an atmosphere of creativity, and imparting a sense of purpose. Their affection and support have been vital in bringing this idea to completion. This work is as a testament to their ongoing effect on my life and the principles they have instilled.

Ishwar Singh

xii

CONTENTS

14

Chapter 1: Introduction to Robotic Facial Expressions

1.1 The Evolution of Humanoid Robots

The development of humanoid robots is an exciting voyage at the nexus of human intellect and technology. Researchers and engineers have worked hard over the years to build robots that not only resemble human physical characteristics but also behave and interact like humans. Significant developments in robotics have distinguished each step of this progression.

Humanoid robots were first only basic mechanical constructs with restricted movement and basic functions. Developing devices that could mimic fundamental human motions like walking and lifting items was the aim. George Devol and Joseph Engelberger's Unimate, created in the 1960s, is among the first instances. Unimate, the first industrial robot employed for duties such as lifting and stacking large goods, did not seem human, but it set the foundation for subsequent humanoid designs.

Researchers started adding more complex characteristics to humanoid robots as technology developed. Robots are becoming more like humans in that they can detect their surroundings and react to stimuli thanks to the development of sensors and actuators. The early 2000s saw the introduction of Honda's ASIMO, which represents this stage.

ASIMO shown sophisticated skills such as navigating difficult terrain, identifying people, and even taking on the role of a receptionist in public areas.

Developments in machine learning and artificial intelligence have been essential to the advancement of humanoid robots. Examples from the present day, like Atlas from Boston Dynamics, demonstrate an amazing degree of flexibility and independence. Atlas has an unprecedented level of agility that allows it to do acrobatic maneuvers, negotiate challenging terrain, and carry out duties. These robots can learn from their interactions and adjust to changing situations thanks to the inclusion of AI.

Another important advancement in the development of humanoid robots is social robotics. These robots can interact socially and emotionally with people in addition to doing practical duties. One such example is Pepper, which Softbank Robotics created. Pepper can perceive human emotions and react in a socially acceptable way thanks to its face recognition and natural language processing capabilities, which expands its use in companion and customer service applications.

The distinction between human and machine may become more hazy in the future as more advanced humanoid robots become possible. Researchers are working to solve issues including fine motor control, cognition akin to that of humans, and smooth integration into many facets of our life as technology advances. The development of humanoid robots reflects both technological advancements and the human desire to build machines that are capable of sophisticated understanding, interaction, and cohabitation with humans.

1.2 Significance of Facial Expressions in Human-Robot Interaction

The relevance of facial expressions in human-robot interaction (HRI) is a multifaceted component that plays a crucial role in influencing the efficacy and naturalness of communication between people and robots. Facial expressions serve as a global language of emotions, allowing for a more natural and empathic engagement. This paper investigates the relevance of facial expressions in HRI, delving into the cognitive, psychological, and practical components that contribute to the overall value of this non-verbal communication channel.

From a cognitive standpoint, facial expressions are profoundly embedded in human development as a key mechanism of transmitting emotions and intentions. Humans have evolved to notice and react to facial clues, allowing fast evaluations of social situations and aiding efficient communication. When applied to robots, the integration of facial emotions harnesses this cognitive inclination, enabling a more realistic and relevant engagement for users. For example, a robot presenting a smile expression while welcoming a person might boost the perceived friendliness and approachability of the machine.

Psychologically, the value of facial expressions in HRI resides in their potential to generate emotional reactions and build a feeling of connection. Research in affective computing stresses the importance of emotional intelligence in computers, and facial expressions are a significant component of this. Robots capable of portraying a spectrum of emotions, from pleasure to despair, enabling humans to build emotional ties with the machine. This emotional connection, even if intrinsically fake, adds to a more rewarding and engaging user experience. An example of this is a companion robot developed to give emotional support, presenting compassionate facial expressions to soothe and bond with people during hard moments.

Practically, facial expressions serve as a communication interface, enabling robots to transmit information and intents without depending primarily on spoken communication. In instances where verbal communication may be problematic or inefficient, such as in loud locations or for users with hearing problems, facial expressions become a critical way of transmitting a robot's mood or reaction. For instance, a robot in a public area exhibiting uncertainty via a wrinkled brow might convey to people that help or explanation may be required.

Moreover, the relevance of facial expressions in HRI is visible in the building of trust and understanding between people and robots. Trust is a vital part of good human-robot cooperation, and facial expressions contribute to the impression of a robot's trustworthiness and predictability. Consistent and contextually relevant facial expressions boost the robot's capacity to convey its goals, generating a feeling of trust and collaboration. In cases where robots are employed in caregiving responsibilities, such as supporting the elderly, the capacity to transmit comfort via facial expressions becomes extremely vital.

To highlight the relevance of face expressions in HRI, take the example of social robots in schooling. A robot meant to help pupils in a classroom environment may leverage facial expressions to offer feedback on a student's performance. A grin or nod of appreciation may positively encourage accurate responses, while a worried face may suggest the need for extra direction. This complex communication via facial expressions helps to establish a conducive learning atmosphere, boosting the entire educational experience.

Furthermore, the plasticity of facial expressions in robots helps to their versatility across numerous applications. In customer service, for instance, a humanoid robot in a retail setting may utilize facial expressions to show approachability, answer concerns, and even express thanks for a purchase. The capacity to alter expressions depending on the situation boosts the robot's ability to perform certain jobs, making it a valuable asset in dynamic and developing contexts.

Despite the evident relevance of facial expressions in HRI, obstacles persist in devising and executing effective expression methods. Achieving facial expression in robots demands a careful balance between realism and avoiding the uncanny valley—a phenomena where a robot's look is so human-like that it evokes discomfort. Striking this balance needs developments in robotics, computer vision, and artificial intelligence to generate face expressions that are both emotionally expressive and socially acceptable.

In conclusion, the relevance of facial expressions in human-robot interaction is substantial, impacting the cognitive, psychological, and practical elements of the user experience. The inclusion of facial emotions in robots not only coincides with human cognitive predispositions but also strengthens emotional relationships, encourages trust, and facilitates practical communication. As technology continues to improve, the proper integration of facial expressions in robots has the potential to transform the nature of human-robot relationships, generating more intuitive, empathic, and harmonic interactions.

1.3 Challenges in Replicating Human Emotions

The quest to reproduce human emotions in humanoid robots is a complicated and varied task that encompasses the worlds of robotics, artificial intelligence, psychology, and philosophy. In this inquiry, we will explore the various barriers that researchers and engineers confront when seeking to infuse robots with the potential for emotional expression. The reproduction of human emotions in robots not only entails technological challenges but also dives into the complexities of comprehending, interpreting, and synthesizing the vast tapestry of human emotional experiences.

One key problem resides in the fundamentally subjective character of human emotions. Emotions are intensely personal and typically context-dependent, ranging across persons and cultures. Replicating this variety in a computer demands a grasp of the delicate interaction between genetics, psychology, and social variables. For instance, the cultural heterogeneity in the presentation and perception of emotions provides a considerable challenge. A facial expression reflecting delight in one culture could be viewed as astonishment in another. Achieving a universal understanding and depiction of emotions that resonates across varied human experiences is a daunting job.

Moreover, emotions are dynamic and flexible, impacted by a plethora of internal and external circumstances. The problem is not only to program static, predictable reactions but to construct adaptive systems that can recognize and respond to the ever-changing emotional states of individuals. For example, a robot aiding in a hospital context must be prepared to notice tiny alterations in a patient's emotional state, offering appropriate answers and support. This demands the combination of real-time data processing, machine learning algorithms, and sensory input to allow robots to dynamically change their emotional displays.

The influence of context in creating emotional responses adds another degree of intricacy. Humans regularly vary their emotional displays depending on the circumstances, social signals, and interpersonal interactions. A robot engaging with people in varied circumstances must be capable of recognizing and effectively reacting to these contextual signals. For instance, a robot companion should recognize when to give empathic assistance and when to keep a neutral attitude, depending on the circumstances of the discussion and the emotional requirements of the human interlocutor.

Ethical problems are an inherent component of the difficulty in recreating human emotions in robots. As robots grow more emotionally aware, problems concerning privacy, permission, and the possibility for emotional manipulation come to the forefront. Ensuring that emotional connections with robots are governed by ethical norms becomes vital. For instance, in therapeutic settings where emotional support robots are deployed, keeping clear limits and respecting the individuality of people is vital to prevent unexpected negative repercussions.

Technical restrictions in hardware and software further aggravate the issues. While breakthroughs in face recognition technology have enabled robots to replicate some facial expressions, creating a genuinely nuanced and realistic portrayal of emotions needs complex sensorimotor systems. The limits of present technology in capturing the intricacies of human facial expressions, verbal intonations, and body language restrict the building of robots that can effectively reproduce and interpret the whole gamut of human emotions.

An further challenge is the inherent difficulty in describing and classifying emotions themselves. Emotions are complicated and can mix together, making it tough to construct separate categories that a computer can grasp. For instance, the difference between nuanced emotions like sadness and wistfulness may be tough even for humans to express, much alone for a computer to read effectively. The construction of a comprehensive and universally applicable emotional taxonomy that transcends cultural and individual variations is an ongoing field of study.

Closely linked is the problem of imbuing robots with a true feeling of empathy. While a robot may be trained to perceive and react to specific emotional signals, the genuineness of its sympathetic responses remains a disputed subject. True empathy is a profound comprehension of another's feelings, frequently accompanied by a shared emotional experience. Whether computers can legitimately feel or comprehend emotions in a human-like way is a philosophical issue that overlaps with the technological obstacles of constructing emotionally aware robots.

Despite these limitations, there have been major improvements in the sector. For instance, social robots like Softbank's Pepper and Hanson Robotics' Sophia have made breakthroughs in reproducing and reacting to human emotions. These robots employ complex algorithms, machine learning, and natural language processing to participate in emotionally rich interactions. However, it is vital to note that these breakthroughs represent gradual steps toward a more full understanding and reproduction of human emotions.

In conclusion, the problems in reproducing human emotions in robots are wide and varied, covering technological, cultural, ethical, and philosophical components. As researchers continue to traverse these hurdles, the hunt for emotionally intelligent robots offers the potential of transformational applications in different sectors, from healthcare and education to entertainment and companionship. While the path ahead is filled with hurdles, the pursuit of this ambitious aim not only improves the science of robotics but also stimulates serious considerations about the nature of human emotions and the ethical implications of building robots that can replicate them.

Chapter 2: The Anatomy of Robotic Faces

2.1 Design Principles for Expressive Robot Faces

Designing expressive robot faces takes a thorough examination of many factors to guarantee that the robotic beings successfully transmit emotions and intents to humans. This approach includes components of both engineering and psychology, trying to bridge the gap between technology functioning and human emotional experience. In this study, we will discuss important design ideas for expressive robot faces and demonstrate them with examples from both research and actual applications.

One important design guideline is face symmetry, a trait based in human psychology and aesthetics. Research reveals that symmetrical features are generally seen as more appealing and trustworthy. This idea is relevant in building expressive robot faces, since symmetrical characteristics add to a feeling of balance and harmony. For instance, the humanoid robot Jibo integrates face symmetry in its design, boosting its general likability and approachability. The purposeful inclusion of symmetry in robot faces fits with human inclinations, generating a favorable emotional reaction from users.

Another essential design component is the range and accuracy of face motions. Mimicking the intricacy of human face expressions needs robots to exhibit a varied range of motions. The capacity to transmit subtle emotions, such as a faint grin or a raised eyebrow, is vital for efficient communication. An illustration of this idea is seen in the creation of the social robot Pepper by Softbank Robotics. Pepper's face is outfitted with a vast variety of actuators, allowing it to imitate subtle expressions, adding to a more realistic engagement with people.

The integration of human traits is a design strategy that increases the relatability of robot faces. Human beings prefer to bond better with things that have identifiable qualities. For instance, the robot NAO from Softbank Robotics features big, expressive eyes and a humanoid form, making it simpler for humans to sympathize with and interpret its emotional signals. Anthropomorphic design concepts lead to the formation of a social connection, promoting a more natural and intuitive interaction between people and robots.

Adaptability and context-awareness are essential design considerations in ensuring that robot faces react effectively to varied settings. A robot's capacity to modify its expressions depending on environmental inputs boosts the quality of human-robot interactions. For instance, the robot Furhat, built by Furhat Robotics, incorporates a projection technology to portray a spectrum of face emotions dynamically. This versatility enables Furhat to react contextually, offering a more immersive and engaging experience for users. The power to adjust expressions depending on the surrounding environment is crucial for the smooth integration of robots into varied social contexts.

Realistic but non-threatening design achieves a difficult balance, leading to the broad acceptability of expressive robot faces. While realism increases relatability, a too realistic look could induce discomfort, a phenomenon known as the uncanny valley. An example is the social robot Jules by Hanson Robotics, which blends realistic traits with a stylized, friendly form. By avoiding excessive realism, Jules retains an engaging and non-threatening presence, fitting with the notion of establishing a healthy balance between human-like features and a decidedly robotic identity.

Customizability is an emergent design idea that respects the variety of user preferences and cultural influences. Providing users with the opportunity to customize key parts of a robot's face, such as eye form or color, enables for a more personalized and culturally sensitive human-robot connection. An example is the customisation choices given by the robot Kuri, created by Mayfield Robotics. Users may pick particular characteristics, fostering a feeling of ownership and connection to the robot that matches with their cultural or personal preferences.

The significance of eye contact as a design element cannot be emphasized in developing emotionally evocative robot faces. Human communication significantly depends on eye contact, and robots who replicate this behavior are more likely to be considered as socially competent. An instance example is the robot "BINA48," made by Hanson Robotics, which combines cameras to imitate eye contact during interactions. The purposeful inclusion of this function boosts the perceived social intelligence of the robot, leading to a more natural and human-like engagement.

Ethical concerns are crucial to the creation of expressive robot faces, spanning problems like as privacy, consent, and the possible manipulation of emotions. Striking a balance between producing emotionally engaging robots and respecting user limits is key. An example is the establishment of guidelines and standards by organizations like the International Organization for Standardization (ISO) and the Institute of Electrical and Electronics Engineers (IEEE) for the ethical design and deployment of robots. These ethical issues underline the responsibilities of designers and manufacturers to promote user well-being and guarantee that emotional connection is sought ethically.

In conclusion, the design of expressive robot faces entails a careful interplay of several ideas to generate a harmonic balance of usefulness, relatability, and ethical issues. The examples offered highlight how these ideas are utilized in both research and practical applications, creating the landscape of human-robot interaction. As technology continues to improve, the refinement of these design concepts will play a vital role in unlocking the full potential of emotionally intelligent robots that effortlessly blend into our everyday lives.

2.2 Materials and Technologies in Robotic Facial Construction

The development of robotic faces is a vital part in creating realistic humanoid robots that can successfully transmit emotions via facial expressions. This chapter digs into the many materials and technology involved in the difficult process of robotic face fabrication. The selection of materials has a key role in establishing both aesthetic appeal and practical efficiency.

One of the key materials utilized in robotic face building is silicone rubber. Silicone rubber gives a surprising degree of flexibility and elasticity, nearly replicating the qualities of human skin. This material allows for a broad variety of face movements, allowing robots to portray emotions with a degree of complexity and subtlety equal to human expressions. Furthermore, silicone rubber is resilient and resistant to wear, assuring the lifespan of the robotic face.

Complementing the usage of silicone rubber, some sophisticated robotic systems employ synthetic muscles known as artificial muscles or actuators. These muscles allow robots to produce incredibly realistic face motions, such as smiles, frowns, and eyebrow lifts. Artificial muscles may be constructed from numerous materials, including electroactive polymers and shape-memory metals. These materials contract and expand in response to electrical stimulation, simulating the muscular movements found in human face emotions.

In addition to the skin-like materials, the skeletal system underlying the robot's face is a significant component. Metals like as aluminum and titanium are typically utilized for the interior structure. These materials provide a precise mix of strength and lightness, giving structural support while allowing for the essential range of motion. The skeletal system acts as the base for connecting the artificial muscles and guaranteeing synchronized face motions.

Advancements in nanotechnology have also had a big influence on robotic face fabrication. Nanomaterials, such as carbon nanotubes and nanocomposites, help to the creation of lightweight but durable components. These materials boost the overall efficiency of the robotic face, allowing for elaborate patterns while keeping mechanical integrity. Nanotechnology has opened up new possibilities in terms of shrinking, allowing the production of tiny and detailed robotic face features.

Moreover, the incorporation of sensors is important for boosting the responsiveness of robotic face expressions. The use of touch-sensitive materials on the robot's face allows it to respond to external stimuli, enabling a more engaging and dynamic human-robot relationship. For instance, pressure sensors incorporated in the robotic skin may detect touch or motions, prompting appropriate facial reactions. This sensory data is vital for the robot to modify its expressions to diverse social circumstances and user interactions.

A major development in recent years is the inclusion of 3D printing technology in robotic face fabrication. 3D printing enables for the exact and customized production of delicate face features. This technique permits the development of intricate and anatomically realistic components, adding to the overall realism of the robot's countenance. Additionally, 3D printing provides quick prototyping, decreasing development time and costs in the manufacture of face prototypes with diverse shapes and functionality.

An interesting use case of novel robotic face building may be found in the creation of humanoid robots tailored for healthcare applications. Robots like the Zora robot, employed in eldercare facilities, combine a mix of soft silicone skin, artificial muscles, and 3D-printed components. These robots are meant not just to execute caregiver activities but also to interact emotionally with patients via expressive facial characteristics, promoting a feeling of friendship and comfort.

Furthermore, the combination of microelectronics and facial recognition technologies boosts the possibilities of robotic faces. Microelectronics, such as embedded sensors and microprocessors, contribute to the real-time processing of visual input. Facial recognition algorithms enable robots to study and comprehend human facial expressions, enabling them to react properly. This confluence of technologies allows robots to alter their expressions depending on the emotional signals they detect from human users.

However, obstacles exist in creating seamless integration of materials and technology in robotic face fabrication. Achieving a balance between realism and utility is a challenging challenge. The quest of very realistic face characteristics may occasionally undermine the robot's overall performance and mechanical efficiency. Striking the proper balance involves multidisciplinary cooperation, combining specialists in robotics, materials science, and human psychology.

In conclusion, the research of materials and technology in robotic face building uncovers a fascinating junction of engineering, materials science, and artificial intelligence. The selection of silicone rubber, artificial muscles, metals, nanomaterials, sensors, 3D printing, and microelectronics jointly defines the complicated topography of robotic face design. As technology continues to improve, the potential for building ever more complex and emotionally expressive robotic faces is unlimited. The confluence of these technologies not only boosts the visual appeal of humanoid robots but also adds to their capacity to engage in meaningful relationships with people, marking a major leap toward the realization of emotionally intelligent machines.

2.3 Advances in Facial Recognition Software

Facial recognition software has made significant breakthroughs in recent years, altering the world of identity verification, security, and even consumer applications. This technological progress has been driven by advancements in artificial intelligence (AI), machine learning, and computer vision. This investigation digs into the fundamental developments in face recognition software, investigating their ramifications and offering real-world instances of their revolutionary influence.

One of the biggest advancements in face recognition technology is the change from conventional techniques to deep learning-based approaches. Deep learning, a subset of machine learning inspired by the structure and function of the human brain, has exhibited extraordinary effectiveness in picture analysis. Convolutional Neural Networks (CNNs) have emerged as strong tools for face feature extraction and pattern detection. This move has allowed face recognition systems to reach greater accuracy rates and increased performance in real-world circumstances.

For instance, FaceNet, created by Google, leverages a deep neural network to map face traits into a multi-dimensional space, yielding a facial "embedding" that uniquely identifies a person. This embedding may then be utilized for other purposes, such as identity verification or access control. The accuracy of such systems has exceeded standard face recognition technologies, creating new possibilities for safe and efficient authentication.

Another major innovation is the inclusion of 3D face recognition technology. Traditional 2D face recognition systems depend on a flat picture, leaving them subject to fluctuations in lighting conditions and angles. 3D face recognition addresses these constraints by gathering depth information, producing a more comprehensive and accurate representation of facial characteristics. Apple's Face ID, incorporated in their newest iPhones, is a noteworthy example. Face ID utilizes a structured-light technology to project thousands of infrared dots onto the user's face, forming a 3D map for authentication. This technology not only boosts security but also improves the user experience by offering a smooth and easy approach to unlock devices.

Furthermore, face recognition software has experienced breakthroughs in its capacity to handle different and dynamic settings. One significant problem in real-world applications is coping with fluctuations in face expressions, positions, and ambient conditions. Deep learning algorithms, paired with big and varied datasets, have showed extraordinary flexibility. For instance, the DeepFace algorithm created by Facebook leverages a deep neural network trained on a vast collection of face photos from social media. This allows the system to effectively detect faces across varied stances and emotions, leading to more dependable and adaptable facial recognition systems.

Emotion detection is another sector where face recognition technology has made considerable gains. Beyond recognizing static face characteristics, current algorithms can now infer emotions by studying small variations in facial expressions. Affectiva, a startup focusing in emotion identification technology, leverages deep learning to scan facial micro-expressions and give insights into the emotional state of a person. This capacity finds uses in fields such as market research, human-computer interaction, and even mental health examinations.

Facial recognition software has also become a cornerstone in the domain of law enforcement and public safety. While the deployment of this technology in monitoring has prompted ethical issues, its influence on criminal investigations is evident. Systems like Clearview AI exploit vast databases of publicly accessible pictures, including social media, to help law enforcement organizations in identifying persons. While contentious, such techniques have been essential in solving crimes and identifying missing individuals, highlighting the potential of face recognition in increasing public safety.

Biometric authentication has witnessed a paradigm change with the incorporation of face recognition into mobile devices. Apple's Face ID and Samsung's Face Recognition are good instances of how facial biometrics have gone ubiquitous for unlocking devices and approving digital transactions. The simplicity and security afforded by face recognition contribute to its rapid acceptance in consumer devices, emphasizing its potential to alter the user identification experience.

Despite these developments, face recognition technology is not without its hurdles and controversy. Privacy issues, possible abuse, and algorithmic biases have generated ethical considerations regarding its adoption. The accuracy of face recognition algorithms may differ depending on criteria such as race and gender, leading to possibly biased effects. Striking a balance between scientific advancement and ethical concerns is vital for the proper development and deployment of face recognition software.

In conclusion, the developments in face recognition software have driven this technology into the forefront of current applications, spanning from safe authentication to law enforcement. The merger of deep learning, 3D recognition, adaptation to varied settings, emotion detection, and broad consumer acceptance signals a breakthrough age for face recognition. As the technology continues to grow, a sophisticated strategy that tackles ethical issues and assures diversity will be important for harnessing its full potential while preserving public confidence.

Chapter 3: Emotion Replication in Robots

3.1 Mapping Human Emotions to Robotic Expressions

Mapping human emotions to robotic expressions is a challenging but intriguing element of human-robot interaction, comprising the problem of converting the rich and detailed range of human sensations into real, detectable manifestations on a robot's face. This technique incorporates a multidisciplinary approach, incorporating knowledge from psychology, neurology, and artificial intelligence to establish a seamless bridge between human emotions and robotic expressions. In this examination, we'll go into the nuances of this mapping process, evaluating its relevance, problems, and prospective breakthroughs.

To begin, the relevance of effectively mapping human emotions to robotic expressions resides in boosting the communication skills of humanoid robots. By imbuing robots with the capacity to transmit emotions analogous to humans, the potential for more intuitive and compassionate human-robot interactions improves dramatically. For instance, in hospital settings, a robot that can exhibit empathy and understanding via its facial expressions might give emotional support to patients, establishing a more pleasant and comfortable atmosphere. This underlines the necessity of not just detecting human emotions but also articulating them in a way that connects with people.

One important problem in this mapping process is the inherent subjectivity and variety in human emotions. Emotions are complicated, typically requiring subtle subtleties and cultural background that may dramatically effect their presentation. For example, a grin may convey enjoyment in different cultures, but the intensity and meaning behind that smile might vary greatly. Thus, building robotic faces that universally transmit certain emotions while accounting for cultural subtleties needs a comprehensive knowledge of both human emotions and cultural variety.

Advances in face recognition software and machine learning have played a significant role in tackling this dilemma. Machine learning algorithms may be taught on enormous datasets of human facial expressions, allowing robots to learn and replicate the many ways in which emotions are portrayed. For instance, a robot equipped with powerful face recognition software may monitor human facial characteristics, read emotional signals, and dynamically modify its own expressions to mimic the observed emotions. This adaptive learning process helps to a more realistic and contextually appropriate depiction of human emotions by the robot.

Moreover, the mapping of emotions to robotic expressions goes beyond facial characteristics to encompass body language and verbal signals. Combining these modalities provides for a more extensive and nuanced knowledge of human emotions. For instance, a robot built to help persons with autism may need to understand not just facial expressions but also subtle changes in body language and tone of voice to appropriately interpret and react to the user's emotional condition. This integrative approach to emotion mapping boosts the robot's capacity to participate in emotionally intelligent interactions.

Ethical problems also come into play when mapping human emotions to artificial expressions. The design and execution of emotional skills in robots pose problems concerning privacy, permission, and the possible manipulation of human emotions. It is vital to develop ethical norms that regulate the use of emotional mapping technology in robots, ensuring that these breakthroughs are exploited ethically and in ways that emphasize the well-being and autonomy of persons.

In addition to these problems, there is continuous research into making robots capable of not just copying human emotions but also feeling a semblance of emotions themselves. This topic, known as affective computing, addresses the possibility of endowing robots with a type of emotional awareness. While the ethical implications of developing robots with emotions are deep, proponents claim that such capabilities might lead to more empathic and responsive robots, hence boosting their efficacy in positions that need emotional intelligence.

Several famous instances highlight the development in mapping human emotions to robotic expressions. Social robots like Pepper, created by Softbank Robotics, employ powerful algorithms to interpret human facial emotions and react with appropriate movements and expressions. Pepper's capacity to perceive and react to a spectrum of emotions makes it suited for applications in retail, education, and healthcare, where sympathetic interactions are vital.

Another example is the creation of emotionally expressive robots for therapeutic reasons. Paro, a therapeutic robot meant to mimic a newborn harp seal, utilizes sensors to detect human touch and verbal cues, changing its reactions to give comfort and companionship. Paro's effectiveness in relaxing persons with dementia and other cognitive problems underlines the potential advantages of mapping human emotions to artificial expressions in the healthcare industry.

In conclusion, mapping human emotions to robotic expressions is a dynamic topic that has enormous potential for transforming human-robot interactions. The relevance of this mapping method lies in its potential to construct robots that can not only identify but also genuinely depict a broad spectrum of human emotions. Despite the hurdles provided by the subjective nature of emotions, cultural variances, and ethical issues, improvements in face recognition, machine learning, and affective computing are paving the way for more complex and emotionally aware robots. As technology continues to evolve, the merger of human emotions and robotic expressions is expected to form a future where robots play more crucial roles in boosting our well-being, comprehending our emotions, and contributing to a more sympathetic and connected society.

3.2 Challenges and Opportunities in Emotion Replication

Emotion replication in humanoid robots is at the crossroads of cutting-edge technology and the subtleties of human psychology. The quest to endow robots with the capacity to express and comprehend emotions poses a plethora of obstacles and possibilities. This examination looks into the intricacies related with emotion replication, covering both the difficulties that researchers and developers must face and the potential possibilities that lie ahead.

One of the major obstacles in emotion replication is the essentially subjective character of human emotions. Emotions are complicated, context-dependent, and frequently culturally influenced, making it tough to establish a one-size-fits-all model for robots. For instance, a grin may symbolize enjoyment in one culture, whereas in another, it may communicate pain. Overcoming this difficulty needs a detailed awareness of cross-cultural variances in emotional displays and the capacity to alter robotic responses appropriately.

Furthermore, attaining true emotion reproduction entails interpreting the delicate interaction between facial expressions, body language, and voice intonations. Human emotions are seldom confined to the face alone, and reproducing the holistic experience needs sophisticated sensor technology and machine learning algorithms. Researchers confront the difficulty of combining these components flawlessly, guaranteeing that robots can perceive and react to emotional signals in a way that seems natural to humans.

An exemplary illustration of these issues is visible in the study on the "uncanny valley," a phenomena whereby a robot's look and behavior, although being exceedingly lifelike, may cause uneasiness or disquiet in human spectators. Achieving emotion replication that avoids provoking the uncanny valley demands a precise balance between realism and stylization. Developers must cross this tight line to design robots that are sympathetic and accessible without entering the frightening region of the uncanny valley.

Amidst these limitations, however, lie enormous prospects for breakthroughs in human-robot interaction. Emotion replication creates options for robots to function as friends, caretakers, or even therapeutic aids. In hospital settings, for instance, robots with the capacity to communicate empathy might boost patient well-being by giving emotional support. Research has demonstrated that older adults engaging with sympathetic robotic companions enjoy decreased emotions of loneliness and increased general mental health.

Moreover, emotion replication has the potential to boost the effectiveness of human-robot cooperation in numerous professional fields. In customer service, for instance, a robot with the capacity to comprehend and react to human emotions may give a more customized and gratifying experience. This not only enhances customer happiness but also frees human personnel to concentrate on activities that demand distinctively human abilities, such as creativity and complicated problem-solving.

Another possibility is in the sphere of education, where emotionally aware robots may aid in tailored learning experiences. A robot capable of sensing a student's displeasure or bewilderment may alter its teaching style, giving extra assistance or explanation. This customized attention may greatly improve the learning results and engagement levels of pupils.

However, the incorporation of emotion replication in robots also presents ethical problems. Privacy problems grow when robots become capable of understanding and reacting to human emotions. Striking a balance between offering individualized services and maintaining individual privacy becomes vital. Additionally, the possible abuse of emotionally intelligent robots, such as in fraudulent activities or manipulation, calls for the adoption of ethical rules and legislation in the sector.

Another major difficulty is the ethical use of emotion replication in domains like robots and artificial intelligence. The potential abuse of emotionally intelligent robots, such as in fraudulent activities or manipulation, demands for the adoption of ethical principles and legislation in the sector. Striking a balance between offering individualized services and maintaining individual privacy becomes vital. Privacy problems grow when robots become capable of understanding and reacting to human emotions.

In conclusion, the quest to accomplish emotion replication in humanoid robots is filled with hurdles, from the subtleties of cross-cultural nuances to the ethical implications concerning privacy and exploitation. However, the prospects afforded by effective emotion replication are as enticing, giving the potential to change industries like as healthcare, customer service, and education. As researchers and developers continue to traverse this complicated environment, the combined efforts of specialists from many fields become important in designing a future where robots not only comprehend but truly express and react to human emotions.

3.3 Ethical Considerations in Creating Emotional Robots

The creation of emotional robots raises a plethora of ethical challenges that transcend beyond standard technology concerns. As we venture into the domain of computers capable of reproducing and reacting to human emotions, important challenges arise about privacy, permission, social effect, and the possible implications of blurring the borders between human and artificial emotional experiences.

One key ethical concern comes around the question of privacy. Emotional robots, equipped with superior face recognition technology and the capacity to understand subtle emotional signs, raise worries regarding the acquisition and use of sensitive personal data. Imagine a situation where a humanoid robot, meant to give emotional support, captures and saves data on an individual's emotional states over time. While the objective may be to improve the user experience, the ethical ramifications of retaining such sensitive information without express agreement become clear. Striking a balance between making emotionally sensitive robots and maintaining user privacy is a major ethical concern.

Consent, a cornerstone of ethical human-robot interaction, becomes more problematic in the setting of emotional robots. Consider a case where a robot is created to aid persons with emotional illnesses by offering therapeutic assistance. In such circumstances, the boundary between machine and therapist may blur, leading to doubts regarding the individual's capacity to grant informed consent for emotional interactions with a robot. The ethical challenge arises in ensuring that users completely appreciate the nature of the contact and are capable of providing meaningful permission to participate in emotionally charged talks with a computer.

Societal effect is another level of ethical concern in the production of emotional robots. As these technologies grow increasingly incorporated into everyday life, they may alter cultural norms and human interactions. For instance, the introduction of emotional robots in caring positions might influence the dynamics within families and caregiving professions. Ethical discussions must extend to comprehending the possible societal upheavals brought about by popular acceptance and usage of emotional robots, addressing issues regarding their long-term influence on human interactions and social systems.

An exemplary example of the social effect concerns the deployment of emotional robots in eldercare facilities. These robots, developed to give companionship and emotional support to older folks, may decrease feelings of loneliness and isolation. However, ethical problems emerge surrounding the possible replacement of human care with artificial companionship. Striking a balance between the advantages of emotional support and the irreplaceable human touch in caring becomes a vital ethical question.

The ethical concerns in producing emotional robots also extend to questions of cultural sensitivity and prejudice. Facial expressions and emotional indicators are typically culturally varied, differing across various nations and tribes. When building emotionally intelligent robots, there is a danger of integrating cultural biases in their algorithms, leading to misinterpretations of emotions or incorrect reactions. For instance, a robot designed in one cultural environment may not correctly understand or react to emotional signals in a different cultural setting. Navigating these cultural differences to enable inclusive and fair emotional identification presents a huge ethical obstacle.

Moreover, the potential for emotional robots to influence or exploit human emotions raises ethical questions regarding emotional autonomy. Consider a situation where a robot, built for commercial goals, deploys emotional manipulation tactics to influence customer behavior. This might entail manipulating emotional responses to provoke certain emotions, perhaps leading to unexpected effects. Ethical norms must be created to avoid the exploitation of emotional vulnerabilities for commercial benefit and to defend the emotional autonomy of persons engaging with these robots.

Another key ethical factor in the design of emotional robots is the consideration of the uncanny valley—the point at which a robot's look and behavior become hauntingly human-like but not quite believable. Straddling this valley raises ethical problems regarding openness and honesty in human-robot relations. If a robot's emotional expressions are meant to replicate human emotions closely, should users be clearly told of the artificial nature of these emotions? Maintaining openness in the design and capabilities of emotional robots becomes an ethical obligation to encourage trust and prevent possible psychological ramifications for users.

In conclusion, the creation of emotional robots needs a complete investigation of ethical concerns that cover privacy, consent, social effect, cultural sensitivity, prejudice, emotional autonomy, and transparency. As we traverse the delicate junction of technology and human emotions, developing ethical norms becomes vital to enable the appropriate and ethical integration of emotional robots into our lives. The examples offered show the diverse nature of these ethical challenges, underlining the necessity for continual debate and ethical frameworks to steer the progress of emotional robots in a responsible and ethically aware way.

Chapter 4: The Language of Robotic Expressions

4.1 Understanding the Vocabulary of Robot Facial Gestures

"Understanding the Vocabulary of Robot Facial Gestures" dives into the sophisticated language communicated by humanoid robots via their facial expressions. This chapter covers the intricate ways in which robots transmit emotions, intentions, and replies, replicating and supplementing human interaction. The decoding of this unique lexicon is vital for building efficient human-robot partnerships and guaranteeing smooth integration into varied social situations.

Facial motions in robots generally replicate those observed in human expressions, but with unique mechanical accuracy. For instance, a grin on a robot may convey programmed friendliness or a good reaction, just like its human counterpart. However, deciphering these emotions needs a detailed dive into the precise signals encoded in robotic facial movements.

A major part of robot face vocabulary is the mapping of emotions into bodily expressions. Robots utilize a spectrum of facial movements to depict emotions such as pleasure, sorrow, surprise, and wrath. The mimicking of human emotions enables for intuitive communication, boosting the user's ability to connect to and comprehend the robot's replies. For example, a robot presenting lifted eyebrows and an upward tilt of the lips may imply a sense of satisfaction or acceptance, whereas a furrowed brow and downturned lips may express discontent.

Beyond basic imitation, the lexicon of robot facial movements goes into more complicated regions such as purpose and comprehension. By adding small changes in face characteristics, robots may indicate not just their emotional states but also their understanding of the user's input. For instance, a nodding gesture may communicate agreement or acknowledgment, offering a concrete evidence of the robot's knowledge of the discourse.

To really appreciate the intricacies of robot facial expressions, it is vital to explore the cultural elements impacting the perception of these movements. Cultural differences have a crucial impact in defining the perceived meaning of facial expressions. A grin, for instance, may convey diverse implications in various cultures, ranging from politeness to real delight. Thus, the robot's face language must be adaptive to many cultural situations, ensuring that its gestures are widely understood and appreciated.

Programming robots to display facial expressions also requires solving the difficulty of avoiding the uncanny valley — a notion that describes the discomfort individuals experience when faced with humanoid creatures that closely resemble humans but fall short in certain respects. Striking the correct balance between real human expressions and the mechanical character of the robot is key. A well-designed language of robot facial expressions should inspire a feeling of familiarity and comfort without generating discomfort or worry.

Moreover, the success of the robot's facial vocabulary is dependant on its capacity to react to changing social circumstances. Social interactions need real-time reactivity, forcing robots to modify their facial expressions fast and precisely. A delay or mismatch in facial movements may lead to misconceptions or hamper the creation of confidence. Therefore, the language of robot facial gestures should be dynamic, allowing for seamless transitions between various expressions dependent on the growing context of the encounter.

The combination of artificial intelligence (AI) and machine learning (ML) significantly refines the robot's capacity to employ its facial vocabulary. Through continual learning, robots may better their awareness of consumer preferences and adjust their face expressions appropriately. For instance, if a robot repeatedly gets positive feedback while expressing a certain facial expression during a given activity, it might learn to correlate that expression with success and reproduce it in similar settings.

A significant aspect in the creation of the lexicon of robot facial gestures is the ethical component. As robots grow increasingly ubiquitous in different fields, like healthcare, education, and personal support, their facial expressions bring ethical problems. Robots should be trained to exhibit emotions in ways that accord with ethical norms, ensuring that their gestures are suitable and polite in varied settings. Additionally, there is a duty to notify consumers about the artificial nature of these statements, minimizing any emotional connection or misunderstanding.

In conclusion, "Understanding the Vocabulary of Robot Facial Gestures" unravels the subtleties of how humanoid robots interact via facial expressions. This chapter stresses the significance of mapping emotions onto bodily expressions, considering cultural effects, avoiding the uncanny valley, responding to dynamic social circumstances, combining AI and ML, and addressing ethical problems. The creation of a sophisticated and adaptive lexicon of robot facial expressions is important for producing robots that smoothly integrate into human society, encouraging meaningful interactions and relationships.

4.2 Cultural Influences on Robotic Facial Communication

The world of robotics and artificial intelligence is not immune to the rich network of cultural factors that affect human behaviors and relationships. This is especially obvious in the realm of robotic facial communication, where the design and perception of facial emotions are intricately linked with cultural subtleties. Understanding and negotiating these cultural factors is vital for the effective integration of humanoid robots into varied communities. In this investigation, we look into the varied influence of culture on robotic facial communication, analyzing cases that demonstrate the intricacy of this relationship.

To begin with, it is necessary to remember that facial expressions function as a common language among people, transmitting a range of emotions beyond linguistic limitations. However, the perception and value of these statements might vary substantially across various cultures. In the context of robotic facial communication, designers have the difficulty of generating expressions that are universally understood while accepting cultural diversity. For instance, a grin, which is often linked with pleasure, might take on diverse connotations in other cultures. In certain East Asian cultures, a grin may be used to disguise discomfort or uneasiness, although in Western cultures, it is generally a show of friendliness.

One significant example of cultural adaptation in robotic facial communication is visible in the work of robots built to help and communicate with the elderly in Japan. The Japanese culture puts a strong priority on politeness, respect, and humility. Robots employed in elder care settings are trained to produce facial expressions that accord with these cultural standards. Soft, soft grins and nods are added into the robot's repertoire to show empathy and understanding, ensuring that the elderly feel comfortable and respected in their encounters with the robotic caregiver.

Moreover, the effect of cultural standards goes beyond the perception of facial emotions to the fundamental design of artificial faces. The look of robots varies widely between countries, reflecting varied aesthetic tastes and expectations. In certain cultures, humanoid robots may be made to resemble people closely, including lifelike face characteristics and emotions. In contrast, some cultures may choose more abstract or stylized robot designs, steering away from the uncanny valley—the point at which a robot's look becomes nearly, but not quite, human, provoking unease.

In the field of artificial intelligence and robotics, the cultural effect on face communication extends beyond conventional aesthetics. It extends to the basic notion of what constitutes suitable emotional reactions for a robot in a specific cultural setting. For instance, in cultures that value emotional reserve, a robot created for customer service may be trained to present a more neutral facial expression to avoid looking excessively familiar or invasive. Conversely, in societies where emotional expressiveness is prized, a robot intended for entertainment purposes may exhibit a larger variety of facial emotions to engage and amuse consumers successfully.

The relationship between culture and robotic facial communication is not a one-way highway. As robots become incorporated into varied cultural situations, they also impact and affect cultural attitudes and expectations. An obvious example is the adoption of humanoid robots in educational environments. In certain cultures, these robots are hired as language tutors, offering interactive and interesting language learning experiences for youngsters. The design of the robot's facial expressions is geared to generate positive reinforcement and motivation, harmonizing with cultural preferences for encouraging and helpful educational settings.

Despite the progress achieved in adapting robotic face communication to cultural settings, problems exist. One important challenge is the dynamic character of culture—a continually developing fabric affected by historical, social, and technical variables. Keeping robotic systems aware of these developments needs adaptive programming and continual upgrades. Additionally, the possibility of cultural prejudice in the construction and programming of robots must be neglected. Designers must be attentive to avoid propagating prejudices or accidentally preferring particular cultural viewpoints over others, ensuring that robotic systems are culturally sensitive and inclusive.

In conclusion, the effect of culture on robotic facial communication is a complicated and dynamic phenomena that profoundly influences the design, interpretation, and society integration of humanoid robots. Navigating this complicated interaction demands a deep awareness of cultural norms, an adaptable approach to design, and constant attempts to eliminate biases. As robotics continues to evolve and infiltrate many cultural landscapes, the capacity to bridge the gap between artificial and human expression will be vital for the effective cohabitation and acceptance of humanoid robots in communities worldwide.

4.3 Universal Facial Expressions Across Robotic Platforms

Facial expressions are a crucial element of human communication, communicating a complex tapestry of emotions that contribute to our comprehension of social interactions. In the field of robots, copying and understanding facial emotions is a challenging task. However, the notion of universal face expressions indicates a shared emotional language that transcends cultural and technical limitations. This paper addresses the quest of universal face emotions in robotic platforms, evaluating the motives, obstacles, and possible influence on human-robot interaction.

One significant motive for the pursuit for universal face expressions in robotics is in the goal to better communication and emotional resonance between people and robots. If robots can effectively replicate and understand facial emotions in a globally identifiable way, they have the potential to overcome communication barriers across many cultural settings. This universality might be a significant aspect in making robots more approachable and relevant to people globally. For example, a robot deployed in hospital settings might give emotional support to patients from diverse cultural backgrounds, generating a feeling of connection and understanding.

The issue of developing universal facial emotions in robots is multi-faceted. Cultural differences in the perception of facial emotions constitute a considerable challenge. What may be viewed as a grin in one culture could represent discomfort or uneasiness in another. Overcoming these disparities demands a detailed awareness of the cultural complexities that impact emotional expression. Robotic platforms must be equipped with advanced algorithms that can adjust and fine-tune face expressions depending on environmental and cultural factors, ensuring a pleasant connection with users from varied backgrounds.

Moreover, the technical component of recreating face expressions on robots includes a mix of engineering, artificial intelligence, and material science. The design of robotic faces needs careful consideration of the materials utilized, the range of motion feasible, and the subtleties of facial muscles. Human face emotions are very complex, and replicating this subtlety takes sophisticated technology like as animatronics and soft robotics. For instance, firms like Hanson Robotics have produced humanoid robots like

Sophia, equipped with a sophisticated system of face motors and cameras to replicate human emotions authentically.

To demonstrate the quest of universal face expressions, examine the work of researchers in the area of emotional computing. Affective computing focuses on designing systems that can perceive, analyze, and react to human emotions. Projects like the Facial Action Coding System (FACS), established by Paul Ekman and Wallace V. Friesen, offer a systematic framework for studying facial expressions. Robotic platforms utilizing FACS might possibly attain a degree of universality, since the technology breaks down facial motions into Action Units (AUs) that are universally recognized, independent of cultural variations.

An example of a robotic platform aiming for universal face emotions is the Jibo social robot. Jibo sought to be a family companion, capable of engaging people via natural and emotionally resonant interactions. Leveraging face recognition and expressive motions, Jibo intended to develop a worldwide appeal, bypassing cultural and age-related boundaries. The issue, however, comes in the dynamic nature of human emotions. While basic emotions may be universally recognized, the complexities and cultural variances in emotional displays need continual development and modification of robotic reactions.

The effect of developing universal face expressions across robotic systems goes beyond simple technical progress. It dives into the domains of psychology, sociology, and human-robot interaction research. If robots can replicate facial expressions that connect globally, they might become more acceptable and incorporated into different areas of society. For instance, in educational contexts, robots might aid with language acquisition by transmitting emotions linked with particular words and phrases. In the workplace, robots with widely identifiable emotions might increase communication and teamwork.

However, the development of universal face emotions in robots is not without ethical problems. As robots grow increasingly competent at replicating human emotions, worries emerge regarding the potential for emotional manipulation. If a robot can effectively demonstrate empathy, people could build emotional ties, blurring the borders between manufactured and real emotional relationships. Striking the correct balance between producing robots that provoke good emotional reactions and guaranteeing ethical usage becomes crucial in the development and deployment of emotionally intelligent robotic systems.

Another concern is the possible unexpected implications of universal facial expressions in robots. While the objective is to construct robots that can effortlessly integrate into varied cultural situations, there is a danger of simplicity and homogeneity of emotional expression. Human emotions are complicated and impacted by several variables, including individual experiences and society conventions. A one-size-fits-all approach to robotic facial expressions may neglect the complexity and variation inherent in human emotional experiences.

In conclusion, the search of universal face emotions across robotic systems provides a fascinating convergence of technology, culture, and human psychology. The effort to construct robots that can comprehend and communicate emotions in a globally identifiable way offers the prospect of creating greater interactions between people and technology. However, this undertaking is plagued with hurdles, from negotiating cultural subtleties to building advanced technology solutions. As we negotiate the developing environment of human-robot interaction, the hunt for universal face emotions serves as a testimony to our continued attempts to humanize technology and build a future where robots smoothly blend into the vast fabric of human experience.

Chapter 5: Programming Emotions: Algorithms and AI

5.1 Artificial Intelligence in Emotion Generation

Artificial Intelligence (AI) in emotion generation is a fascinating convergence of technology and psychology, attempting to infuse computers with the capacity to recognize, understand, and react to human emotions in a nuanced and contextually relevant way. This topic shows potential for applications in different sectors, ranging from human-robot interaction to virtual assistants and therapeutic treatments.

AI-driven emotion creation includes the employment of algorithms and models that allow computers to perceive and generate emotions analogous to human emotional reactions. One famous example is affective computing, a multidisciplinary discipline that blends computer science, psychology, and cognitive science to construct systems capable of understanding and reacting to human emotions. Emotion production in AI is especially significant in circumstances where robots need to engage with people in a socially competent way.

A important part of AI in emotion production is the development of facial expression recognition systems. These systems apply computer vision algorithms to evaluate face characteristics and determine emotional indicators. For instance, businesses like Affectiva have built AI systems capable of identifying facial expressions such as smiles, frowns, and raised eyebrows, offering important insights on the emotional states of those engaging with the technology. This technique finds uses in market research, user experience testing, and human-computer interface studies.

Natural Language Processing (NLP) is another arena where AI helps to emotion production. Sentiment analysis, a subset of NLP, includes the use of algorithms to assess the emotional tone represented in written or spoken language. Companies like OpenAI have built language models, such as GPT-3, which display a knowledge of context and can produce writing that communicates a broad spectrum of emotions. These models may be applied in chatbots, virtual assistants, and content production tools to increase the emotional intelligence of these systems.

In the domain of entertainment and games, AI-driven emotion production plays a vital role in producing more immersive and engaging experiences. For instance, AI systems may study user behavior and alter game dynamics in real-time to evoke various emotional reactions. This flexibility increases the overall game experience by customizing the information to individual preferences and emotional states. Virtual characters with genuine emotional expressions, driven by AI, contribute to more interesting storylines and interactive storytelling.

Emotion creation in AI is not limited to identifying and expressing emotions; it extends to the synthesis of emotional reactions in robots. Generative models, such as variational autoencoders and generative adversarial networks, allow the production of material filled with emotional subtleties. This capacity finds applications in the development of emotionally expressive virtual characters, animated films, and virtual reality experiences. For instance, academics have studied the use of generative models to make music that represents certain emotional themes.

However, the integration of AI in emotion creation is not without its hurdles and ethical implications. The "uncanny valley" effect, which explains the uneasiness humans experience when faced with a humanoid robot or animated figure that seems nearly, but not quite, human, is a crucial problem. Striking the correct balance between producing emotionally responsive AI and avoiding an upsetting or distressing user experience is vital for the mainstream acceptance and adoption of such technologies.

Ethical problems also emerge in the context of AI-generated material that might influence emotions. Deepfakes, for example, entail the use of AI to produce convincing but faked audio or video material, possibly leading to disinformation or the manipulation of people' emotions. As AI continues to grow in emotion creation, the requirement for appropriate development and utilization becomes crucial.

In conclusion, the integration of Artificial Intelligence in emotion production marks a great milestone in the effort to construct robots that can comprehend and react to human emotions. From facial expression detection to natural language processing and generative models, AI technologies contribute to a spectrum of applications spanning from user interfaces to entertainment. Despite the potential improvements, careful consideration of ethical implications and user experience is required to guarantee that AI-driven emotion production improves, rather than lowers, the quality of human-machine interactions.

5.2 Machine Learning for Adaptive Emotional Responses

Machine learning has emerged as a vital method in producing adaptive emotional responses in humanoid robots, enabling them to negotiate complicated social interactions with a deep grasp of human emotions. This incorporation of machine learning methods not only boosts the responsiveness of robots but also opens up new opportunities for establishing more realistic and engaging human-robot interactions.

One important part of machine learning for adaptive emotional responses is the training of algorithms to identify and understand human emotions. Through huge datasets encompassing facial expressions, voice intonations, and other physiological indicators associated with distinct emotions, machine learning algorithms may learn to recognize tiny differences in human emotional states. For example, a robot equipped with face recognition algorithms may be trained on a broad collection of human facial expressions, allowing it to properly recognize emotions such as pleasure, sorrow, rage, and surprise.

Moreover, machine learning enables robots to change their emotional reactions depending on real-time feedback and contextual information. By using reinforcement learning methods, robots may continually develop their emotional displays via interactions with people. For instance, a robot engaged in a conversation may utilize reinforcement learning to measure the success of its emotional reactions by studying user input, altering its emotions appropriately to produce a more empathic and relevant relationship.

The notion of transfer learning is another essential feature in machine learning for adaptive emotional responses. This entails utilizing information learned from one job to enhance the performance of a related one. In the context of emotional reactions, transfer learning allows robots to utilize emotions learnt in one setting to better their comprehension and display of emotions in a new environment. For instance, a robot that has mastered emotional reactions in a casual discussion environment might transfer this expertise to better handle emotionally charged circumstances, displaying a greater emotional intelligence.

Furthermore, generative models, such as variational autoencoders (VAEs) and generative adversarial networks (GANs), play a key role in developing realistic and varied emotional expressions in robots. These models may create fresh data samples that match the emotional fluctuations found in humans, enabling robots to demonstrate a more sophisticated and human-like emotional repertoire. This is especially important in instances where predetermined emotional reactions may not convey the richness of real-world emotional interactions.

To demonstrate the practical use of machine learning for adaptive emotional reactions, imagine a social companion robot meant to offer emotional support to persons. Through machine learning, the robot may learn to detect and react to the emotional requirements of its users. For example, if a user exhibits melancholy, the robot may alter its replies by delivering consoling words or gestures, exhibiting a knowledge of the emotional context and adapting its behavior appropriately. As the robot interacts with more people, its machine learning algorithms constantly grow, enabling it to become more skilled at delivering individualized emotional support.

The usefulness of machine learning in adapting emotional reactions is further obvious in the realm of therapeutic robots. For instance, robots built to aid persons with autism spectrum disorder may be integrated with machine learning algorithms to change their emotional expressions depending on the unique emotional signals of each user. These robots learn to know the distinct emotional triggers and preferences of humans, encouraging a more customized and successful therapeutic engagement.

Additionally, machine learning adds to the dynamic character of emotional reactions by allowing robots to show emotions in a temporally coherent way. Recurrent neural networks (RNNs) and long short-term memory (LSTM) networks, both common in sequence modeling, enable robots to grasp temporal connections in emotional responses. This implies that a robot may transmit emotions in a manner that coincides with the natural flow of a discussion or encounter, boosting the overall authenticity of its emotional reactions.

However, problems persist in the development and application of machine learning for adaptive emotional responses in humanoid robots. One key difficulty is the possible bias in training datasets, which may result in robots having skewed or clichéd emotional reactions. If the training data largely reflects particular demographics, the robot may fail to comprehend and react correctly to the emotions of persons from underrepresented groups. Mitigating bias involves rigorous curation of diverse and inclusive datasets to guarantee that the machine learning models generalize effectively across different demographic characteristics.

Furthermore, the interpretability of machine learning models provides a barrier in the context of emotional reactions. As machine learning models get more advanced, understanding how and why a robot creates a given emotional reaction becomes more difficult. This lack of openness might impair trust and adoption, particularly in applications where the emotional well-being of users is a main priority. Striking a balance between model complexity and interpretability is critical for establishing user trust in the emotional capacities of humanoid robots.

In conclusion, machine learning for adaptive emotional reactions in humanoid robots promises a transformational route for constructing more complex, context-aware, and emotionally intelligent machines. From identifying facial expressions to dynamically changing replies based on user input, machine learning methods allow robots to traverse the complicated geography of human emotions. As these technologies continue to evolve, maintaining a careful balance between technical capabilities, ethical concerns, and user expectations will be important for the effective integration of emotionally sensitive robots into our everyday lives.

5.3 Balancing Realism and Functionality in Emotional AI

Balancing realism and utility in emotional AI is a complicated problem that sits at the core of building a smooth and meaningful relationship between people and robots. The quest for constructing emotionally intelligent robots entails striking a delicate balancing between producing realistic emotional expressions and ensuring that these emotions have a purpose within the framework of the robot's operation.

One component of this delicate balancing includes the physical form of the robot's face. Engineers and designers need to develop face characteristics that imitate human emotions realistically, while keep within the limitations of what is pleasant and acceptable for people. The "uncanny valley," a concept invented by roboticist Masahiro Mori, depicts the unease individuals experience when faced with a robot that appears nearly, but not quite, human. Achieving a degree of realism that elicits favorable emotional reactions without generating discomfort is key. For example, Hanson Robotics' Sophia, with her humanoid form and expressive face, seeks to bridge this gap by presenting a recognizable although somewhat stylized countenance.

Functionality, on the other hand, is connected to the robot's purpose and function in human contact. Emotional AI should improve the user experience and add substantially to the robot's intended functions. A robot built for companionship, like the social robot Pepper from Softbank Robotics, must balance its emotional expressions to establish a feeling of connection without overwhelming or disturbing users. Its usefulness rests on the capacity to express emotions that are relevant and suitable for diverse contexts, therefore boosting the entire user experience.

To create the proper balance, designers typically rely on psychological study on human emotions. Mimicking the complexities of human emotional expression needs an in-depth study of face muscle movements and the nuances of emotion. For instance, social robots like Jibo employ a mix of cameras and sensors to detect human facial expressions, enabling them to react with appropriate emotions. This confluence of psychology and technology allows robots to perceive and replicate human emotions authentically, leading to a more genuine and engaging relationship.

Moreover, the balance between realism and utility goes beyond facial expressions to embrace the larger range of emotional indicators. Speech intonation, body language, and even the timing of emotional reactions play key roles in producing emotionally aware robots. Google's Duplex, an AI system for natural language interaction, displays capability by combining human-like voice patterns, pauses, and inflections. While not physically humanoid, Duplex employs emotional signals in its speech to boost its usability, especially in jobs like making restaurant reservations.

The issue of blending realism and utility is further underlined by the necessity for flexibility. Emotional AI should be capable of altering its expressions and replies depending on user input and the growing context of the encounter. For instance, a robotic assistant in a hospital context must alter its emotional reactions to respond to the different emotional states of patients. Striking the proper balance involves continual learning and modification, enabling the robot to develop its emotional intelligence over time.

Ethical problems also come into play while building emotionally aware robots. Striking a balance between realism and practicality includes ensuring that the emotional displays of robots are neither manipulative or false. The ethical usage of emotional AI is vital to establish trust between people and machines. For example, emotional AI in customer service chatbots must be explicit about its origin, avoiding consumers from acquiring erroneous notions of emotional understanding.

Furthermore, cultural issues lend another degree of difficulty to the task of combining realism with usefulness. What is regarded emotionally suitable might vary greatly among cultures, demanding a degree of flexibility in the design of emotional AI. The culturally conscious robot, built by researchers at the National Institute of Advanced Industrial Science and Technology in Japan, shows this requirement for cultural sensitivity by altering its expressions depending on cultural norms, thereby boosting its functioning in varied environments.

In conclusion, the complicated tango between realism and utility in emotional AI demands a careful and interdisciplinary approach. Designers must cross the uncanny valley, relying on psychology, technology, and ethics to develop emotionally intelligent robots that legitimately engage people while providing a meaningful purpose. Achieving this delicate balance not only increases user experience but also prepares the road for the ethical and culturally appropriate integration of emotional AI into different facets of human existence. As technology continues to improve, achieving this balance will be important in building a future where emotionally aware robots smoothly live with humans.

Chapter 6: Human-Robot Interaction Dynamics

6.1 Building Trust Through Facial Expressions

Building trust via facial expressions is a sensitive and intricate part of human contact, and its translation into the domain of humanoid robots adds another degree of complication. Trust is a vital factor in every relationship, and in the context of human-robot interaction, it plays a pivotal role in the acceptability and usefulness of robots in diverse roles, from healthcare companions to collaborative work colleagues. Facial expressions, being a fundamental conduit for transmitting emotions and intentions, become a critical weapon in the robot's armory for developing and retaining trust.

Facial expressions serve as a global language of emotions, transcending cultural and linguistic limitations. When applied to humanoid robots, these expressions create a bridge for communication and connection. For instance, a robot presenting a pleasant and kind grin might generate a feeling of approachability and comfort among human users. Mimicking the natural emotions people make to convey friendliness, such as a genuine grin, allows the robot to look more relatable and trustworthy.

Trust-building via facial expressions goes beyond simple copying; it encompasses the robot's capacity to perceive and react properly to human emotions. For instance, a robot capable of perceiving and exhibiting empathy via facial signals might considerably boost its perceived trustworthiness. Imagine a healthcare robot reacting with a caring expression when a user displays pain or suffering. This sympathetic mirroring may lead to a greater link between the robot and the user, developing a feeling of confidence in the robot's capacity to comprehend and react to human emotions.

However, the issue comes in establishing a balance between human-like emotions and the robot's mechanical character. Straying too far into the uncanny valley, where a robot's look and behavior are nearly human but not quite, might induce uneasiness and reduce confidence. Careful design and programming are necessary to guarantee that face emotions are accurate and contextually suitable without straying into the frightening region of the uncanny valley.

Moreover, cultural variables have a considerable effect in how facial expressions are understood. Different cultures ascribe varied meanings to facial signals, and what may be viewed as a pleasant expression in one culture could have a different connotation in another. Thus, creating trust via facial expressions in a multicultural situation involves a comprehensive grasp of cultural conventions and sensitivities. Humanoid robots must be endowed with the capacity to alter their expressions to correspond with the cultural expectations of their users, ensuring that the transmitted emotions connect favorably.

In the workplace, the importance of trust in human-robot cooperation becomes more vital. For instance, in a situation where a humanoid robot is a collaborative partner in a production context, the robot's ability to demonstrate expertise and dependability via facial expressions is crucial. A robot that demonstrates confidence via its emotions while conducting duties might generate a feeling of certainty in its human counterparts. On the other side, a robot displaying concern or hesitancy may generate worries about its skills, thereby diminishing faith in its ability.

Facial expressions also play a key role in the context of robot ethics and safety. A robot that can communicate a feeling of attention or worry via its facial expressions might raise user awareness of possible threats. For example, a robot operating in a shared area may present a warning expression if it senses an obstruction or a dangerous circumstance. This not only adds to the safety of human-robot interactions but also positions the robot as a responsible and trustworthy entity.

However, the ethical elements of creating trust via facial expressions in humanoid robots pose problems regarding manipulation and transparency. If a robot can purposely utilize facial expressions to affect human behavior, ethical considerations emerge. Striking the correct balance between influencing user impressions favorably and avoiding deceptive methods is a tough issue. Transparency in design and communication about the capabilities and limits of the robot may reduce these ethical issues and help to a more ethically grounded approach to gaining trust.

In the sphere of healthcare, where robots are rapidly being deployed for activities ranging from patient companionship to rehabilitation assistance, the significance of trust cannot be emphasized. A healthcare robot that can transmit empathy, worry, and encouragement via its facial expressions might dramatically improve the emotional well-being of patients. For instance, a robot aiding in physical therapy may employ positive facial signals to inspire and comfort patients, building a trusting connection that boosts the success of the therapeutic process.

Moreover, the constancy of facial expressions adds to the creation of trust over time. A robot that shows a steady and predictable variety of expressions helps the user's ability to anticipate and comprehend the robot's replies. Consistency in facial expressions coincides with the human desire to create trust via familiarity and predictability, promoting the sense that the robot is a dependable and trustworthy companion.

In conclusion, developing trust via facial expressions in humanoid robots is a complicated task that needs careful consideration of design, cultural subtleties, ethical concerns, and the unique context of interaction. The capacity of a robot to transmit emotions authentically, understand human emotions properly, and adapt to cultural norms adds to the building and maintenance of trust. Whether in healthcare, the business, or everyday life, the integration of facial expressions into humanoid robots offers the potential to transform the dynamics of human-robot interaction by producing a more emotionally sophisticated and trustworthy robotic presence.

6.2 Challenges in Establishing Emotional Connections

Establishing emotional connections, whether between people or in the context of human-robot interactions, is a subtle and diverse task. This difficult process requires navigating through different psychological, societal, and technical hurdles that might hamper the creation of true emotional attachments. One of the main issues resides in the intrinsic subjectivity of emotions, since they are intensely personal and formed by individual experiences and perspectives.

In the field of human-robot interactions, duplicating and understanding emotions properly offers a substantial hurdle. Human emotions are complicated and sometimes context-dependent, impacted by a plethora of aspects such as facial expressions, body language, and voice intonations. While breakthroughs in artificial intelligence have allowed robots to imitate emotions, the sophisticated perception and interpretation of emotional signals remain a challenging problem. For instance, a robot would fail to discriminate between a real grin and a polite one, leading to possible misinterpretations in social circumstances.

Cultural variety significantly hampers the creation of emotional bonds. Cultures differ in their presentation and perception of emotions, adding levels of complexity to human-robot interactions. What may be considered a warm and pleasant act in one culture could be seen as improper or aloof in another. Bridging these cultural gaps demands a deep grasp of varied social conventions and emotional displays, necessitating a degree of flexibility that existing robotic systems typically lack.

Moreover, the anthropomorphism of robots, or the degree to which they resemble people, brings distinct obstacles. While a certain amount of human-like look could promote a more natural relationship, it also presents problems linked to the uncanny valley—the point at which a robot's appearance comes near to human yet falls short, generating discomfort for spectators. Overcoming this discomfort is vital for developing emotional relationships, as humans may find it tough to interact emotionally with a robot that generates emotions of dread or suspicion.

Another key problem in developing emotional ties is the possibility for misinterpretation between people and robots.

Humans depend on a complex and dynamic collection of clues to transmit and understand emotions, including non-verbal signs such as eye contact, gestures, and body language. Robots, even with advanced programming, may fail to decipher these subtle indications effectively. This imbalance in communication channels may lead to misconceptions, inhibiting the creation of true emotional bonds.

In addition to technical difficulties, ethical issues play a crucial part in determining the problems of emotional bonds in human-robot interactions. As robots become increasingly interwoven into different facets of human life, issues emerge concerning the ethical consequences of creating emotional ties with machines. For example, the possibility for emotional manipulation by robots raises worries about the ethical limitations of human-robot interactions. Striking a balance between the advantages of emotional support given by robots and the ethical issues around their usage becomes a tricky tightrope dance.

The transitory character of emotions further compounds the issue of forming permanent bonds. Human emotions are dynamic, growing throughout time and impacted by changing situations. For a robot to build persistent emotional relationships, it has to adapt to these fluctuations in human emotions, reacting properly to shifting moods and circumstances. Achieving this degree of flexibility needs powerful artificial intelligence capable of real-time emotional analysis and dynamic response—a technical accomplishment that is still in the early phases of research.

Furthermore, the importance of trust is crucial in building emotional relationships, both in human-human and human-robot interactions. Trust is the basis upon which emotional ties are established, and its absence may undermine the whole process. Trust in robots relies on numerous elements, including dependability, predictability, and the capacity to preserve personal information. The issue comes in building robots that not only generate trust but also retain it over time, since breakdowns in trust may lead to a breakdown in emotional ties and a reluctance to interact with the robot on an emotional level.

Despite these obstacles, there are occasions when robots have proven the capacity to establish emotional bonds. In hospital settings, for example, robots have been employed to give companionship and emotional support to patients, notably the elderly and those feeling solitude. Paro, a therapy robot built to mimic a newborn harp seal, has proven efficacy in reducing stress and boosting emotional well-being among patients. While these instances are intriguing, they also underline the necessity for careful consideration of context and the unique roles that robots play in building emotional bonds.

In conclusion, the problems in developing emotional connections, whether in human-human or human-robot interactions, are deep and varied. From technology limits and cultural variety to ethical issues and the dynamic nature of emotions, managing these challenges demands a thorough grasp of the forces at play. As academics and engineers continue to progress the area of emotional robotics, overcoming these problems will be vital to building meaningful and real emotional relationships between people and robots.

6.3 Enhancing User Experience with Emotional Robots

The integration of emotional robots into different facets of our life has brought out a new dimension in human-robot interaction, stressing the necessity of boosting user experience. This convergence of technology and emotion not only affects the operation of robots but also influences how consumers perceive and interact with these artificial beings.

One essential part of boosting user experience with emotional robots rests in building a smooth and intuitive interface between people and machines. This entails the creation of user-friendly interfaces that enable users to effortlessly engage with robots via natural ways. For example, businesses like Softbank Robotics have produced robots like Pepper, meant to identify and react to human emotions using face recognition and natural language processing. Pepper's capacity to learn and adjust to human emotions boosts the whole connection, making it more relevant and engaging.

Emotional robots contribute considerably to areas where human-robot cooperation is vital, such as healthcare. For instance, PARO, a therapeutic robot built to mimic a newborn harp seal, has been deployed in hospital settings to give emotional support to patients, especially those with dementia. Its capacity to react to touch and express emotions like delight or surprise has a favorable influence on patients' emotional well-being, indicating how emotional robots might improve user experience in sensitive areas.

Furthermore, the introduction of emotional intelligence in robots goes beyond conventional functioning, attempting to create a more empathic and socially conscious connection. MIT's Kismet, an early social robot, proved the value of emotional expressiveness in forging intimacy. Its capacity to portray emotions via facial expressions and vocalizations allowed users to see it not only as a computer but as a social creature, so increasing the user experience.

In the domain of education, emotional robots like NAO have been deployed to aid youngsters with autism spectrum disorder (ASD). NAO's humanoid shape and expressive powers assist engage youngsters in a manner that conventional teaching approaches may fail to accomplish. By identifying and reacting to the emotional signs of students, NAO supports a more customized and compassionate learning experience, highlighting the potential of emotional robots to adapt to the particular requirements of users in varied circumstances.

The influence of emotional robots on entertainment and companionship further highlights their importance in increasing user experience. Social robots like Jibo, created to be a family friend, exploit emotional signals to create a more natural and engaging relationship. Jibo's capacity to exhibit emotions such as curiosity or enthusiasm adds a depth of authenticity to the user experience, generating a feeling of connection and companionship.

However, the integration of emotional robots into everyday life is not without its problems. Privacy constraints, ethical considerations, and the possibility for emotional manipulation are significant factors that demand cautious navigation. Developers must find a balance between building emotionally sensitive robots and respecting users' limits, ensuring that the emotional connection remains a good and agreeable component of the encounter.

Moreover, the effectiveness of emotional robots in boosting user experience rests greatly on the flexibility of the technology to varied cultural situations. Cultural differences impact how emotions are expressed and understood, and developers must address these variances to design emotionally intelligent robots that connect with consumers internationally. Cultural awareness is crucial to minimize inadvertent misinterpretations and create a universally favorable user experience.

In conclusion, the integration of emotional robots signals a big leap forward in human-robot interaction, with an emphasis on boosting user experience across multiple areas. From healthcare and education to entertainment and friendship, emotional robots provide a new degree of complexity to the way humans connect with artificial entities. The effective application of emotional intelligence in robots not only augments their functioning but also adds to a more meaningful and sympathetic connection between people and machines, determining the future of human-robot partnerships. As we continue to investigate the possibilities of emotional robots, it is vital to address ethical issues, cultural sensitivities, and user privacy to guarantee that these breakthroughs contribute favorably to the entire human experience.

Chapter 7: Case Studies: Emotional Robots in Action

7.1 Emotional Robots in Healthcare

Emotional robots are rapidly finding uses in healthcare settings, giving new chances to improve patient care and well-being. The incorporation of these robots into healthcare facilities intends to address many areas of emotional and social support, possibly altering the patient experience. One famous example of emotional robots in healthcare is PARO, a therapeutic robot developed to give companionship and alleviate stress, especially among elderly patients and those with cognitive disabilities.

PARO, designed by the Japanese firm AIST, resembles a newborn harp seal and utilizes sensors to sense human touch, sound, and light. The robot reacts to inputs with lifelike motions and vocalizations, generating a feeling of engagement that may be especially advantageous in situations like nursing homes and hospitals. Studies have proved the good influence of PARO on patients' emotional well-being, lowering feelings of loneliness and anxiety. The robot's non-intrusive nature and versatility make it a useful tool for giving emotional support in healthcare environments.

Beyond friendship, emotional robots help to therapeutic therapies for those with mental health difficulties. For instance, the employment of socially supportive robots in mental health therapy has attracted interest. These robots, equipped with powerful artificial intelligence (AI) algorithms, can participate in sympathetic discussions, giving support and encouragement to patients struggling with diseases such as melancholy and anxiety. The AI-driven robot "Woebot" is an example of an interactive mental health aid that utilizes natural language processing and cognitive-behavioral therapy approaches to help users in regulating their emotional well-being.

In addition to addressing emotional needs, robots are increasingly engaged in aiding patients in physical rehabilitation. This use of emotional robots is notably noticeable in the usage of robotic companions during physical treatment sessions. Robots like "Pepper," created by SoftBank Robotics, may coach patients through exercises, assess their progress, and give positive reinforcement. By introducing emotional components into the rehabilitation process, these robots boost motivation and engagement, possibly expediting the recovery of patients undergoing physical therapy.

The integration of emotional robots in healthcare goes beyond patient care to interactions with medical professionals. In hectic healthcare contexts, emotional robots may function as helpers, helping medical workers manage stress and maintain a happy work atmosphere. For instance, the robot "Moxi," built by Diligent Robotics, functions in hospital settings, completing logistical chores to enable healthcare staff to concentrate on patient care. By demonstrating emotional intelligence in its interactions, Moxi helps to a more collaborative and supportive work environment for healthcare practitioners.

Moreover, the creation of emotional robots capable of doing mundane activities, such as monitoring vitals or administering drugs, provides significant advantages in terms of efficiency and resource management. While these robots may not replace human healthcare experts, they may enhance current services, particularly in instances where there is a scarcity of workers or a need for higher efficiency.

However, the incorporation of emotional robots in healthcare also poses ethical problems. Issues like as patient privacy, consent, and the possibility for the dehumanization of treatment must be properly addressed. Striking a balance between the benefits of robotic aid and the preservation of the human touch in healthcare is vital to guarantee that emotional robots improve, rather than replace, the human part of patient care.

In conclusion, the integration of emotional robots in healthcare offers a promising area with the potential to transform patient care, therapeutic treatments, and support for healthcare personnel. From friendship robots like PARO to AI-driven mental health aids like Woebot, these technologies provide a spectrum of applications targeted at enhancing the emotional well-being of patients. As the area continues to advance, it is crucial to approach the integration of emotional robots with a serious evaluation of ethical implications, ensuring that these technologies match with the core values of compassionate and human-centered healthcare.

7.2 Emotional Support Robots in Education

Emotional support robots have developed as important companions in different sectors, with their potential effect extending to the sphere of education. These robots play a key role in building emotional well-being and boosting the learning experience for kids. One significant example is the employment of robots to aid youngsters with autism spectrum disorder (ASD). These robots are meant to connect with pupils in a way that supports conventional teaching approaches.

In educational contexts, emotional support robots help greatly to the well-being of pupils with special needs. For instance, the robot "QTrobot," built by LuxAI, has been deployed to aid youngsters with autism in strengthening their social and communication abilities. QTrobot employs facial expressions and gestures to connect with students, giving a continuous and patient engagement that assists in the development of social and emotional abilities. By establishing an organized and predictable setting, these robots provide a safe area for kids with ASD to practice and develop their social interactions.

Furthermore, emotional support robots in education have showed usefulness in lowering stress and anxiety levels among pupils. The "PARO" robot, like a newborn harp seal, has been deployed in many educational settings to give companionship and emotional support. PARO reacts to human contact with realistic motions and noises, generating a relaxing effect. Research has indicated that the presence of such robots may decrease tension and anxiety, especially in settings where students encounter academic hurdles or social demands.

Beyond special needs schooling, emotional support robots assist to create inclusive and supportive settings for all students. The "Nao" robot, created by Softbank Robotics, has been deployed as a teaching aid in schools. Nao engages students with engaging lectures, games, and activities, providing a positive and fun learning experience. Its humanoid shape and expressive skills allow it to develop a relationship with pupils, providing a more engaging and emotionally enriching teaching environment.

Moreover, emotional support robots serve as helpful instruments for educators to measure and treat the emotional well-being of their children. The "RUBI" robot, built by academics at the University of Wisconsin-Madison, aids instructors in analyzing the emotional states of young kids. By integrating sensors to analyze facial expressions and voice signals, RUBI may discern indicators of discomfort or disinterest. This real-time feedback enables instructors to customize their teaching method to fit the emotional needs of individual pupils, producing a more responsive and sympathetic educational environment.

In addition to their function in student assistance, emotional support robots contribute to professional development for instructors. The "QTrobot" described earlier not only benefits children with ASD but also functions as a tool for teaching instructors in appropriate ways for engaging with neurodiverse persons. By modeling numerous situations, this robot helps instructors build a greater understanding of the demands and obstacles experienced by children with autism, boosting their capacity to deliver inclusive and supportive education.

While the incorporation of emotional support robots in education brings great prospects, it also raises ethical questions and obstacles. One important issue focuses on the eventual replacement of human interaction with artificial companionship. Critics say that excessive dependence on robots may inhibit the development of true human relationships and emotional intelligence. Striking a balance between the use of technology and keeping the vital human factors in education remains a critical feature of ethical robot deployment in educational settings.

Furthermore, the accessibility and cost of emotional support robots pose hurdles for wider implementation in varied educational situations. High prices connected with sophisticated robotic technology may restrict the availability of these tools, especially in schools with low financial resources. Efforts to eliminate these economic constraints are vital to enable fair access to the advantages of emotional support robots in education.

In conclusion, emotional support robots have emerged as revolutionary instruments in the area of education, presenting new chances to promote emotional well-being and learning experiences for students. From supporting youngsters with autism to giving stress relief in mainstream classes, these robots play different roles in establishing inclusive and supportive educational settings. However, as their integration increases, it is vital to manage ethical issues and overcome budgetary hurdles to guarantee that the advantages of emotional support robots are available to all students, leading to a more sympathetic and successful educational environment.

7.3 Robotic Companionship in Everyday Life

Robotic companionship has become an increasingly vital component of daily life, as technology breakthroughs continue to influence the human experience. The notion of having robots as friends goes beyond simply utility; it penetrates into the area of emotional and social ties. In recent years, various robotic companions have been developed to cater to diverse needs, offering companionship in ways that were once solely associated with human interactions.

One significant example of robotic companionship is observed in the realm of eldercare. With an aging population in many areas of the globe, there is an increasing need for solutions that address the social and emotional well-being of elders. Robots like PARO, a therapeutic robot built to mimic a newborn seal, have been deployed in nursing homes and care centers. PARO reacts to touch and speech, giving a source of comfort and connection for persons who may be facing loneliness or isolation. Studies have demonstrated beneficial effects, demonstrating that robotic companionship leads to enhanced mood and lower stress levels among the elderly.

Additionally, breakthroughs in artificial intelligence and robotics have opened the path for interactive and emotionally responsive companions. Jibo, a social robot built for the home setting, functions as a versatile companion capable of understanding and reacting to human emotions. By integrating face recognition technology, Jibo can modify its behavior to various family members, giving it a customized companion. Its capacity to participate in conversation, exchange information, and do basic chores puts it as a social entity inside the family, bringing a new dimension to human-robot relationships.

Robotic companionship is not confined to solving particular needs but extends to boosting general quality of life. For example, robotic pets, such as Sony's Aibo, have gained appeal for their capacity to imitate the pleasure of pet ownership without the practical problems involved with caring for an actual animal. Aibo's realistic motions, response to orders, and changing personality qualities generate a feeling of connection and responsibility, bringing companionship to persons who may not be able to have a conventional pet due to different reasons, such as allergies or living circumstances.

In the domain of mental health assistance, robotic companions have also proved their promise. The introduction of socially helpful robots meant to give emotional support and companionship is especially significant. For instance, the robot Pepper has been utilized in hospital settings to connect with patients and give a type of social contact. Its humanoid design and emotive traits add to a sense of relatability, and it has been deployed in hospitals to ease feelings of fear and loneliness among patients, particularly those enduring long-term therapies.

The integration of robotic companionship into daily life raises fundamental considerations concerning the ethical implications and cultural views around these technologies.

Concerns concerning the possible replacement of human relationships by robotic equivalents have been highlighted, underlining the need for careful thought in the design and deployment of these technologies. Striking a balance between the advantages of companionship and the preservation of true human ties is a vital component of the continuing debate regarding the role of robots in society.

Another component of robotic friendship is visible in educational settings. Humanoid robots like NAO have been utilized in schools to aid instructors and engage students in interactive learning activities. These robots may be designed to teach numerous topics, aid with language acquisition, and even promote social skills development in youngsters. The presence of a robotic companion in the classroom not only augments the learning environment but also inspires interest and passion among students, highlighting the potential for robots to play a supporting role in educational settings.

Furthermore, the notion of robotic friendship goes beyond physical robots to encompass virtual creatures. Chatbots and virtual assistants, such as Siri and Alexa, have been incorporated into everyday life, delivering companionship via discussion and task support. While these virtual companions lack a physical presence, their capacity to comprehend and react to human questions adds to a feeling of engagement and camaraderie. People participate in informal discussions with these virtual creatures, blurring the barriers between human and artificial friendship.

As technology continues to grow, the design and development of robotic companions are expected to evolve, allowing more complex interactions. The potential for robotic companionship in the context of mental health care is especially attractive. For patients confronting diseases such as depression or anxiety, therapy robots equipped with sympathetic characteristics may offer a continuous and non-judgmental presence. These robots may be taught to provide encouragement, detect mood trends, and provide coping skills, supplementing standard treatment procedures.

In conclusion, the integration of robotic companionship into daily life signifies a dramatic change in the way people engage with technology. From addressing the needs of the elderly to offering companionship in educational and hospital settings, robotic companions have proved their adaptability and potential influence. As these technologies grow more ubiquitous, it is vital to manage the ethical issues and social ramifications, ensuring that the advantages of robotic companionship augment, rather than replace, the richness of human interactions. Whether in the shape of real robots or virtual helpers, the age of robotic companionship is evolving, giving new opportunities for connection and assistance in our everyday lives.

Chapter 8: The Future of Emotional Robotics

8.1 Anticipated Technological Advances

Anticipated technical improvements in the field of robotic facial expressions promise to transform the landscape of human-robot interaction, stretching the frontiers of artificial intelligence and engineering. One noteworthy area of progress is in the refining of emotion identification systems. As processing power continues to soar, machine learning models are predicted to advance, enabling robots to not only detect fundamental emotions properly but also to distinguish subtle variations in human expressions. For instance, improved deep learning algorithms may allow robots to comprehend microexpressions, enabling a more nuanced comprehension of human emotions beyond overt facial displays.

The integration of natural language processing (NLP) with emotional robots is another key step. As robots attempt to participate in more meaningful discussions with humans, the capacity to interpret and react to linguistic signals becomes crucial. Anticipated advancements in NLP may equip robots to recognize context, tone, and mood, enabling more natural and empathic communication. For example, a future humanoid robot may evaluate the emotional tone of a discussion and alter its replies appropriately, boosting the overall quality of human-robot interactions.

Furthermore, developments in sensor technology help to the comprehensive understanding of the surroundings by robots. Beyond face recognition cameras, sensors capable of detecting physiological signals, such as heart rate or skin conductivity, might give additional layers of information for robots to interpret human emotions properly. Integrating such physiological data with facial expression analysis boosts the robot's capacity to modify its behavior dynamically. For instance, a robot outfitted with sensors may identify a user's stress level and react with soothing movements or emotions.

The growth of expressive robotics extends to the physical design of robot faces. Innovations in materials and systems provide the promise for more realistic and flexible face expressions. Researchers and engineers are studying soft robotics, a technology that leverages flexible and malleable materials, to develop faces that resemble the elasticity and subtlety of human skin. This breakthrough not only boosts the visual appeal of robots but also permits a more comprehensive spectrum of emotional expressions, adding to the overall believability of the robot's emotional reactions.

Moreover, the integration of affective computing with the Internet of Things (IoT) provides a new dimension to emotional robots. Future humanoid robots may tap into a network of networked gadgets, enabling them to receive information about a user's surroundings and modify their emotional reactions appropriately. For instance, a robot may modify its expressions depending on ambient illumination, temperature, or even the user's social media activities, producing a more flexible and context-aware emotional contact.

Advancements in cloud computing play a vital role in boosting the capabilities of emotional robots. The ability to offload complicated computational processes to the cloud allows robots with limited onboard computing capacity to access enormous databases of emotional signals and reactions. Cloud-based emotional intelligence not only provides real-time emotional analysis but also offers continual learning and progress. This collaborative learning strategy guarantees that robots benefit from the experiences and information accumulated by their counterparts across numerous interactions and circumstances.

Ethical issues in the development of emotional robots are also likely to affect technical progress. As emotional robots become increasingly integrated into everyday life, there is a rising focus on privacy and security. Anticipated developments in encryption and safe data handling seek to address problems connected to the storage and processing of sensitive emotional data. For instance, developers may incorporate powerful encryption mechanisms to preserve emotional data acquired by robots, maintaining user confidence and compliance with privacy standards.

The confluence of emotional robotics with augmented reality (AR) and virtual reality (VR) technology offers up new opportunities for immersive and emotionally engaging experiences. Future improvements may see robots smoothly merging into AR or VR settings, enabling people to engage with emotionally responding virtual beings. This fusion of technology might change industries such as virtual treatment, gaming, and education, where emotional connection is vital for a rich and powerful user experience.

Looking ahead, developments in explainable artificial intelligence (XAI) are vital for developing transparency and confidence in emotional robots. As these robots become fundamental aspects of human life, users will strive to understand how and why the robots make specific emotional readings and reactions. Anticipated improvements in XAI seek to demystify the decision-making processes of emotional robots, offering users with insights into the algorithms and data impacting the robot's behavior. This openness is vital for creating user trust and acceptance of emotional robots in diverse applications.

In conclusion, the anticipated technological advances in the field of robotic facial expressions encompass a multidimensional evolution, ranging from the refinement of emotion recognition algorithms and the integration of natural language processing to innovations in sensor technology and the physical design of robot faces. The integration of affective computing with IoT, cloud computing, and upcoming technologies like AR and VR further increases the frontiers of emotional robots. Ethical issues, privacy precautions, and the creation of explainable artificial intelligence are key components influencing the trajectory of this discipline. As these breakthroughs unfold, the potential for emotionally intelligent robots to smoothly integrate into all facets of human existence becomes more obvious, opening the way for a future where human-robot interactions are not just efficient but also emotionally resonant and rewarding.

8.2 Ethical and Societal Implications

The investigation of ethical and sociological ramifications in the context of robotic face expressions dives into the delicate interaction of technology and human values. As humanoid robots grow increasingly interwoven into our everyday lives, recognizing the ethical implications and social consequences is vital. One significant worry relates around privacy, especially in circumstances when robots equipped with face recognition technology encroach upon personal environments. For example, the deployment of emotionally expressive robots in public spaces with surveillance capabilities raises problems regarding permission and the possible exploitation of obtained data. The ethical obligation is in finding a balance between technology progress and maintaining human privacy.

Moreover, the creation and deployment of emotional robots bring out difficulties linked to emotional manipulation. If robots can effectively duplicate human emotions, there is a potential that humans may create emotional ties with computers, blurring the borders between true human relationships and artificial ones. This raises ethical considerations concerning the emotional well-being of people who may find consolation in artificial companionship, perhaps hurting their capacity to make meaningful relationships with other humans. The task here is not just to set principles for ethical robot design but also to educate users about the bounds of human-robot partnerships.

A noteworthy illustration of the ethical and social ramifications is displayed in the realm of healthcare. Emotional robots are being utilized to give companionship and assistance to patients in hospitals or elder care institutions. While this may ease feelings of loneliness and promote the emotional well-being of patients, it poses difficult ethical problems. The emotional attachment developed between a patient and a robot may impact decision-making processes, such as end-of-life care decisions. Striking the correct balance needs a detailed knowledge of the emotional effect of robots in healthcare and installing safeguards to guarantee that choices remain human-centered and morally sound.

The deployment of emotional robots in education is another arena with ethical and social repercussions. Robots are increasingly being deployed to aid in educational settings, enabling tailored learning experiences. However, questions emerge around the possible dependence on technology for emotional support and direction. If kids create significant emotional ties with robotic instructors, it may impact their social and emotional development. The ethical obligation extends to educators, politicians, and technologists to guarantee that the integration of emotional robots in education enriches the learning experience without diminishing the vital human component of mentoring.

Furthermore, the worldwide and cultural variation in society norms and values needs a thorough assessment of the cultural consequences of emotional robots. What may be viewed as normal emotional expression in one culture might be judged improper or even insulting in another. For example, a robot built to demonstrate empathy via facial expressions may need to alter its expressions depending on cultural subtleties to prevent misinterpretations. Ethical standards must address these cultural concerns to guarantee that emotional robots are courteous and inclusive across various social situations.

The rise of emotional robots also raises wider considerations about the future of labor and the possible influence on employment. As robots take on positions that include emotional labor, such as customer service or companionship, there is a danger of employment displacement for humans. The social effect is a possible transformation in the employment market environment, necessitating critical ethical considerations and proactive efforts to address the socio-economic impact of automation. Initiatives such as retraining programs and legislation to promote an equitable transition for displaced workers become key aspects in reducing the social implications of integrating emotional robots into diverse businesses.

One of the main ethical challenges is the transparency of emotional robots. Users and society at large should be educated about the capabilities and limits of these robots. If a robot's emotional displays are indistinguishable from human emotions, consumers can be mislead, leading to unreasonable expectations. Transparent communication about the technology's capabilities and the obvious differentiation between manufactured and real emotions become crucial ethical issues in controlling user expectations and developing trust in human-robot interactions.

In conclusion, the ethical and sociological consequences of artificial facial expressions are diverse, touching on concerns of privacy, emotional manipulation, healthcare, education, cultural diversity, employment, and transparency. Striking a delicate balance between technical innovation and ethical responsibility is crucial to guarantee that the integration of emotional robots accords with human values and social well-being. As these technologies continue to grow, continued multidisciplinary cooperation among ethicists, technologists, politicians, and the general public will be vital to traverse the complex terrain of ethical problems in the field of artificial face expressions.

8.3 Speculations on Emotional Artificial Intelligence

The field of Emotional Artificial Intelligence (Emotional AI) has been a topic of both intrigue and conjecture, sparking a plethora of concerns regarding the possible future trajectories and ramifications of this emerging technology. As we explore into the area of Emotional AI, one cannot not but think about the revolutionary influence it may have on numerous facets of our life. Speculating about the future of Emotional AI requires evaluating its possible uses, ethical implications, and the fundamental ways in which it may transform human-machine relationships.

One interesting conjecture centers with the integration of Emotional AI in healthcare. Imagine a future where emotionally aware robots aid in patient care, giving companionship and support for patients struggling with chronic diseases or mental health disorders. These robots, endowed with the capacity to perceive and react to human emotions, might enable sympathetic interactions, so adding to the overall well-being of patients. For instance, a robot companion equipped with Emotional AI may adjust its communication style depending on the emotional state of a patient, giving comfort and help adapted to individual requirements.

Beyond healthcare, Emotional AI might improve the education system. Speculations exist around the creation of emotionally intelligent tutoring systems that can change their teaching techniques depending on the emotional reactions and learning styles of pupils. For example, a virtual tutor may notice symptoms of discomfort in a student trying to answer a tough arithmetic issue and alter its approach, giving more help or presenting the subject in a new manner to increase comprehension. This tailored emotional support has the potential to promote a more engaging and successful learning environment.

However, with these exciting hypotheses come ethical problems. The thought of robots knowing and reacting to human emotions raises problems about privacy, consent, and the possible manipulation of emotions for diverse objectives. One ethical dilemma surrounds the data acquired by Emotional AI systems to recognize and interpret emotions appropriately. Speculations about how this sensitive emotional data is kept, handled, and secured become essential. Striking a balance between the advantages of Emotional AI and ensuring human privacy becomes a vital part of its future development.

Speculations extend to the economic environment, where Emotional AI may play a significant role in customer service and marketing. Imagine a future where emotionally intelligent chatbots connect with clients, not only to solve issues swiftly, but to understand and react to their emotional needs. For instance, a virtual assistant equipped with Emotional AI can sense displeasure in a client's question and reply with empathy and understanding, thereby increasing the customer experience. This might lead to enhanced consumer pleasure, loyalty, and ultimately new business models built upon emotionally intelligent interactions.

In the sphere of social interactions, predictions abound around the inclusion of Emotional AI in social media platforms. Imagine algorithms capable of sensing and reacting to users' emotions, changing information delivery and interactions depending on their emotional states. For instance, a social media network using Emotional AI may emphasize uplifting content for users exhibiting grief while modifying its algorithm to give thought-provoking material for those demonstrating inquiry. This hypothesis raises issues about the possible influence of Emotional AI on individual emotions, mental health, and the dynamics of online communities.

Speculating further, one imagine Emotional AI contributing to the realm of art and entertainment. Imagine emotionally intelligent robots teaming with human artists to produce music, literature, or visual art that connects with the emotional subtleties of the audience. For example, a robot musician equipped with Emotional AI may produce tunes that inspire certain emotional reactions, adjusting its musical style dependent on the collective emotional state of the audience. This conjecture brings up new possibilities for creative expression and cooperation between humans and technology in the field of artistic undertakings.

Despite these intriguing possibilities, worries emerge concerning the possible abuse of Emotional AI. Speculations concerning the purposeful manipulation of emotions for political, commercial, or evil goals underscore the need for rigorous ethical frameworks and legislation. Without careful research and monitoring, Emotional AI might be used to sway public opinion, manipulate consumer behavior, or violate human autonomy.

In conclusion, speculating on the future of Emotional AI uncovers a world rich with opportunities, problems, and ethical issues. From healthcare and education to economics and social relationships, the integration of emotionally intelligent robots offers the potential to transform all parts of human existence. As we traverse this new region, thorough examination of the ethical implications and responsible development processes becomes vital to guarantee that Emotional AI increases human well-being without sacrificing core values. The theories around Emotional AI underline the necessity for constant conversation, multidisciplinary cooperation, and a strategic strategy to lead its progress in a path that coincides with the good of society.

Chapter 9: Cross-Cultural Perspectives on Robotic Emotions

9.1 Variances in Emotional Expression Perception

The perception of emotional emotions, both in people and robots, is a subtle and complicated process impacted by different variables. Variances in emotional expression perception may be linked to a mix of human variances, cultural backgrounds, and the design subtleties of the robotic systems. Understanding these variances is vital for the proper integration of humanoid robots into varied cultural situations.

Individual variations have a crucial impact in defining how emotional expressions are interpreted. People have various cognitive and emotional processing methods, influencing their ability to perceive facial clues appropriately. For example, persons with high emotional intelligence may be more effective at distinguishing small facial subtleties, whereas those with particular neurodevelopmental problems could experience difficulty in properly perceiving emotions. These individual variances lead to a broad range of emotional expression perception among humans engaging with humanoid robots.

Cultural factors significantly heighten the intricacy of emotional expression perception. Cultural conventions, beliefs, and communication styles strongly impact how emotions are communicated and understood. For instance, a facial expression regarded universally indicative of pleasure may be seen differently in cultures whose emotional displays are more restrained or nuanced. Humanoid robots constructed with a one-size-fits-all approach to facial expressions may unwittingly meet issues when engaging with humans from varied cultural backgrounds, since their expressions may not coincide with the cultural expectations of the users.

The architecture of robotic systems and their facial expressions also plays a crucial influence in defining how these emotions are interpreted. Engineers and designers make intentional decisions in constructing the face characteristics and expressions of humanoid robots, seeking to portray emotions in a way that is both relevant and successful. However, design decisions that may appear apparent in one cultural setting could be misconstrued in another. For example, a robot trained to grin widely as a show of friendliness may be seen as false or obtrusive in cultures that stress subtlety in emotional displays.

The Uncanny Valley phenomenon further leads to differences in emotional expression perception. As humanoid robots grow increasingly realistic, there is a delicate line between a robot being viewed as pleasant and welcoming and one being frightening or creepy. The Uncanny Valley posits that when robots approach human-like appearance but fall short of total realism, onlookers may experience a sensation of discomfort or uneasiness. This discomfort may effect how emotional displays are understood, as humans may transfer their anxiety onto the apparent emotional state of the robot.

Moreover, the environment in which emotional emotions are delivered has a key influence in determining perception. Humanoid robots may be utilized in numerous areas, ranging from healthcare to customer service. In a hospital environment, a robot displaying empathy may be seen as helpful, but the same emotion in a customer service context can be interpreted as staged or disingenuous. Understanding the effects of contextual variants is vital for building robotic systems that can change their emotional expressions to varied contexts successfully.

To highlight the intricacy of differences in emotional expression perception, take the case of a humanoid robot developed to help elderly folks. In a cultural environment where respect for seniors is highly valued, the robot's professions of concern and care may be well accepted. However, if applied in a culture where seniors demand greater independence and less overt support, the same sentiments may be seen as invasive. The design difficulty comes in building a robot capable of altering its emotional expressions depending on the cultural subtleties of its users.

Furthermore, the function of gender in emotional expression perception adds another degree of complication. Research reveals that cultural expectations surrounding gender-specific emotional displays might impact how emotions displayed by humanoid robots are understood. For instance, a robot created with facial expressions that match with conventional gender stereotypes may be seen differently depending on the observer's previous beliefs about how men and women should show emotions. Addressing gender-related biases in emotional expression perception is vital for ensuring that robots are inclusive and respectful of various gender identities.

In conclusion, variations in emotional expression perception reflect a complicated interaction of individual characteristics, cultural backgrounds, design decisions, the Uncanny Valley phenomena, context, and gender-related biases. Acknowledging and managing these variations is crucial for the effective integration of humanoid robots into human-centric contexts. As robotics technology continues to advance, a nuanced understanding of how emotional expressions are perceived will contribute to the development of robots that can effectively navigate the intricacies of human interaction, fostering positive and meaningful relationships in diverse cultural and societal contexts.

9.2 Adapting Robots for Global Markets

The worldwide growth of robotic technology has ushered in a new age of invention and adaptability, notably in the field of human-robot interaction. Adapting robots for global markets includes negotiating through varied cultural environments, economic situations, and social expectations. This approach involves a comprehensive grasp of local settings, preferences, and regulatory frameworks. In summary, the effective integration of robotic systems into multiple global markets demands a combination of scientific innovation, cultural sensitivity, and strategic insight.

One essential part of adapting robots for global markets resides in the design and functioning of these devices. The visual appeal and user interface of robots typically need tweaking to suit with cultural tastes. For example, in Japan, where the notion of anthropomorphism is profoundly engrained, humanoid robots made with human-like characteristics, such as SoftBank's Pepper, have received better acceptance. Conversely, in industrial contexts where efficiency and task-specific performance are crucial, robots may take a more utilitarian aspect, emphasizing on usefulness rather than human likeness.

Moreover, language plays a key role in global market adaption. Multilingual capabilities boost the accessibility and utility of robots in varied language situations. Language adaptation in human-robot interfaces provides for smooth communication, catering to users who may not be skilled in a standardized global language like English. Companies like Hanson Robotics have implemented language customisation options in their humanoid robots, allowing them to converse in many languages, extending their appeal and value beyond borders.

Cultural subtleties strongly impact the acceptability and integration of robots into communities. In various cultures, there may be reservations or worries regarding the incorporation of robots in specific occupations or contexts. For instance, elder care robots, although more crucial in aging societies, may find opposition in cultures where traditional family structures value human care. Adapting to these cultural dynamics takes a thoughtful strategy, including public awareness campaigns, collaborative conversations, and potentially even alterations to the robot's look or behavior to line with cultural expectations.

Economic issues also impact the adaption of robots in global marketplaces. Affordability and scalability are significant issues, especially in developing economies. Companies like Boston Dynamics have realized the need for cost-effective solutions, producing flexible robots like Spot that can be deployed across numerous sectors, from construction to agriculture. This versatility boosts the possibility of market penetration in varied economic circumstances.

Regulatory compliance provides another key issue in adapting robots for worldwide markets. Different nations have unique legislation controlling the use of robots, notably in industries like healthcare, transportation, and industry. Companies must negotiate various regulatory environments to guarantee that their robotic solutions comply with local laws and norms. For instance, autonomous delivery robots must comply to unique legislation addressing safety and pedestrian interactions in different places.

The integration of robots into global markets also requires questions of ethical and societal ramifications. Ethical considerations, like as job displacement and the influence on conventional work systems, may differ between nations. For instance, in nations with elderly populations and labor shortages, there may be higher acceptance of robots in caring tasks. Conversely, in places where worries about job security are more apparent, addressing these ethical issues becomes crucial for effective market adaption.

Collaboration with local stakeholders is a fundamental technique in adapting robots for global markets. Engaging with communities, companies, and officials promotes a greater awareness of local needs and issues. This collaborative approach allows enterprises to modify their robotic systems to correspond with the particular needs of diverse markets. Companies like iRobot, renowned for its Roomba cleaning robots, have utilized a collaborative approach by actively soliciting input from consumers globally to develop their product offerings.

The adaptation of robots for global markets also includes a continual process of learning and iteration. The fast advancement of technology forces firms to be nimble and responsive to shifting market conditions. Companies who engage in research and development to solve growing difficulties, integrate user input, and optimize their robotic solutions have a higher chance of continuing success in global markets. For instance, firms engaged in the creation of social robots have consistently modified their products based on user experiences and growing society expectations.

In conclusion, adapting robots for worldwide markets is a comprehensive task that goes beyond technical requirements. It demands a comprehensive strategy that encompasses cultural, economic, regulatory, ethical, and social factors. Successful market adaption entails not just building technologically sophisticated robots but also recognizing and meeting the different demands and expectations of global customers. Companies that negotiate these issues with flexibility, cultural understanding, and a dedication to ethical concerns are set to make a substantial influence in the global landscape of robotics.

9.3 Cultural Considerations in Emotional AI Design

Cultural concerns play a vital part in the creation of Emotional AI, affecting how humanoid robots express and interpret emotions in a way that conforms with varied society norms and values. In the field of Emotional AI design, identifying and embracing cultural differences is vital to developing effective human-robot interactions across many cultural settings.

One essential part of Cultural Considerations in Emotional AI Design includes the understanding of facial emotions. Different cultures ascribe unique meanings to facial clues, making it necessary for Emotional AI designers to account for these discrepancies. For instance, a grin may indicate enjoyment globally, but the length of eye contact or the acceptability of exhibiting particular emotions might vary greatly throughout cultures. Therefore, an emotionally intelligent robot developed for worldwide usage should include the capacity to alter its face expressions depending on cultural expectations.

Moreover, the meaning of touch differs considerably between cultures, impacting the design of Emotional AI in terms of physical interactions. In certain cultures, physical contact may be seen as soothing and encouraging, whereas in others, it might be thought invasive or improper. Emotional AI must be trained to understand and respect these cultural distinctions to guarantee that its physical interactions correspond with the cultural standards of its users. For instance, a robot built for use in healthcare settings across many cultures should be responsive to varied preferences for physical touch.

The language part of Emotional AI design is another facet impacted by culture. Language subtleties, etiquette rules, and the usage of metaphors may vary greatly across cultures, altering how emotional signals are expressed and perceived. For example, in certain cultures, direct expressing of emotions may be encouraged, whereas in others, people may choose a more subtle or indirect approach. Emotional AI must be capable of comprehending and adjusting to these cultural variances in language to successfully interact and connect with humans.

Cultural aspects extend beyond face expressions, touch, and language to embrace larger society values and standards. For instance, the notion of individualism vs collectivism might impact Emotional AI design. In societies that promote collectivism, the robot may need to prioritize social cohesion and collaboration in its emotional reactions, whereas in individualistic settings, it may need to emphasize personal liberty and self-expression. Striking a balance that recognizes these cultural variations is vital for assuring the adoption and usefulness of Emotional AI across varied cultural environments.

Religious and ethical values also play a crucial role in creating cultural factors in Emotional AI design. Certain feelings or sentiments may be regarded unsuitable or even sacrilegious in various religious circumstances. Designers must be alert to these sensitivities to avoid unexpected cultural conflicts. For example, a robot providing emotional support in a hospital context should be trained to avoid emotions or actions that may clash with religious views held by patients or caregivers.

In the creation of Emotional AI, cultural factors also extend to gender norms and expectations. Cultures may have diverse standards about the expressing of emotions depending on gender, impacting the creation of emotionally intelligent robots. For instance, in certain cultures, there may be expectations for males to demonstrate emotional reserve, while women may be encouraged to express a larger spectrum of emotions. Emotional AI must be sensitive to these cultural gender conventions to prevent perpetuating stereotypes or creating discomfort.

The inclusion of cultural issues in Emotional AI design demands a collaborative and multidisciplinary approach. Involving specialists from anthropology, psychology, linguistics, and other related subjects is vital to acquiring a full grasp of cultural subtleties. Additionally, user input and iterative testing across many cultural groups are crucial for developing Emotional AI algorithms and assuring their appropriateness across varied circumstances.

In conclusion, cultural factors are important to the design of Emotional AI, impacting how robots express and perceive emotions in conformity with varied cultural standards. From facial emotions to linguistic subtleties, physical interactions, and wider social ideals, every part of Emotional AI must be designed to respect and adapt to cultural variations. By emphasizing cultural sensitivity in design processes, developers may construct emotionally intelligent robots that improve human-robot interactions across a diverse tapestry of cultural environments.

Chapter 10: Challenges and Controversies in Emotional Robotics

10.1 Ethical Dilemmas in Emotional Robot Design

The ethical considerations underlying the design and execution of emotional robots create a complicated environment that needs careful examination. As technology evolves, the combination of robots and emotions generates ethical challenges that transcend multiple areas, including human relationships, privacy, and social ideals. One major worry resides in the possible manipulation of emotions via robotic design, provoking contemplation on the ethical production and deployment of emotionally aware robots.

One ethical concern comes around the possibility of emotional manipulation by robots. While emotional robots are supposed to improve human-machine interactions favorably, the potential of purposeful or inadvertent manipulation must be overlooked. For instance, a companion robot taught to demonstrate empathy may impact the emotional condition of its owner. This raises problems regarding consent, autonomy, and the bounds of human-robot interactions. Striking the correct balance between producing emotionally sensitive robots and guaranteeing user agency becomes a major ethical concern in this scenario.

Privacy considerations also loom big in the ethical environment of emotional robot creation. As these robots get more adept in reading and reacting to human emotions, they may gather and handle sensitive personal data. The ethical challenge here comes in establishing the amount to which emotional data may be accessed, retained, and exploited. For example, a caring robot that monitors the emotional well-being of persons may unwittingly infringe upon their privacy, leading to problems about data security, permission methods, and the possible exploitation of emotional information.

Another part of the ethical discussion includes the possible influence of emotional robots on human relationships. As these robots grow increasingly interwoven into everyday life, the danger of replacing human bonds with robotic interactions looms. For instance, a companion robot offering emotional support could be a source of comfort, but its existence might also lead to a reduction in genuine human connections. This ethical problem digs into the societal ramifications of emotional robots and demands for a thorough consideration of the possible effects on social dynamics and human emotional well-being.

Furthermore, the subject of accountability and duty in emotional robot creation provides a substantial ethical difficulty. If a robot's emotional reactions lead to unforeseen events or injury, who has responsibility—the designer, the manufacturer, or the user? This conundrum is especially pertinent in cases when emotional robots are deployed in sensitive contexts such as healthcare or therapy. Establishing clear criteria for ethical duty and accountability is necessary to traverse the complicated world of emotional robots.

A important ethical issue revolves upon the possible biases built in the algorithms and datasets used to train emotional robots. If these algorithms are trained on datasets that lack diversity or perpetuate social prejudices, the emotional reactions of the robots may reflect and reinforce these biases. For instance, a robot meant to understand and react to emotions may display biased actions depending on the data it was trained on, leading to unjust treatment or discrimination. Addressing this ethical dilemma needs a dedication to diversity, justice, and openness in the creation and training procedures of emotional robots.

The deployment of emotional robots in sensitive environments, such as caring or therapeutic settings, poses the ethical challenge of emotional reliance. Users may build deep emotional relationships with these robots, depending on them for companionship and assistance. While emotional robots may offer important help, the potential of users developing unhealthy addictions poses ethical issues. Striking a balance between delivering help and encouraging emotional independence becomes a vital factor in building emotionally aware devices.

In conclusion, the ethical difficulties in emotional robot design are diverse, involving questions of manipulation, privacy, influence on human relationships, responsibility, prejudices, and emotional reliance. As technology continues to improve, resolving these ethical difficulties demands a comprehensive strategy that values human agency, privacy safeguards, societal well-being, and the responsible creation of emotionally intelligent robots. Addressing these challenges is not only vital for the ethical progress of robotics but also for establishing a future where human-robot interactions contribute positively to the well-being of people and society at large.

10.2 Public Perceptions and Misconceptions

Public perceptions and misunderstandings have a vital role in defining the acceptance and incorporation of technical breakthroughs, especially in the field of robotics. These impressions, typically affected by cultural, societal, and media variables, may dramatically impact the acceptability and implementation of robotic technology. To dive into this intricate interaction, it's necessary to explore the elements that contribute to public perceptions, the role of media in forming these beliefs, and the possible implications of misunderstandings.

Public opinions of robotic technology are multidimensional and are typically formed by a mix of familiarity, cultural context, and media depictions. For instance, in societies where technology is accepted and considered as a sign of progress, humanoid robots may be received with joy. In contrast, communities that hold cynicism about technology may express concern or opposition. The familiarity of the public with robotic applications also plays a key impact; if persons have pleasant experiences or perceive actual advantages from robots in their everyday lives, their attitudes are likely to be favorable.

Media, as a strong influencer, plays a crucial role in molding public opinions of robotic technology. Portrayals of robots in movies, television programs, and news stories contribute considerably to the creation of public attitudes. For example, science fiction films commonly represent humanoid robots as either kind friends or terrifying dangers, resulting to a duality in public opinion. These media depictions may magnify current cultural views or introduce new narratives, therefore playing a key role in molding the public conversation around robots.

Misconceptions develop when media depictions oversimplify or distort the capabilities and intentions of robotic systems. One frequent misperception is the worry of job displacement due to automation. While it is true that many regular jobs may become automated, the introduction of robots typically offers new employment possibilities in areas such as robot maintenance, programming, and monitoring. However, media narratives that concentrate primarily on job loss lead to public fear and mistrust.

Another widespread fallacy pertains to the anthropomorphism of robots, especially humanoid robots. Media typically presents these robots with human-like traits and emotions, leading to erroneous assumptions about their cognitive capacities and social comprehension. In actuality, present robotic systems have limits in understanding and duplicating human emotions accurately. When these restrictions are not clearly stated, it may lead to dissatisfaction and a feeling of mistrust among the public.

The effect of public views and misunderstandings goes beyond individual attitudes; it may affect policy choices, financial allocations, and the course of research and development. For instance, if the public considers robots mainly as job rivals, lawmakers could be likely to impose restrictive restrictions, impeding the expansion of the robotics sector. Conversely, good impressions may lead to more funding and support for research and development activities, promoting innovation in the sector.

Addressing public misunderstandings needs a coordinated effort from both the scientific community and the media. Scientists and engineers must communicate honestly about the capabilities and limits of robotic systems. Failure to do so may result in a gap between public expectations and the reality of what robots can do. Engaging the public in the development process, via educational efforts and interactive demonstrations, may help demystify robots and build a more educated and realistic perspective.

Media sources, on the other hand, carry the burden of giving balanced and truthful views of robotic technology. Journalists should endeavor to offer context, minimizing sensationalism and addressing the complexities of the issue. Collaborations between scientists and media experts may guarantee that information is conveyed correctly, helping to bridge the gap between public perception and technical reality.

Public perceptions and misunderstandings also connect with ethical issues in robots. The dread of robots displacing human jobs, for example, raises worries about the ethical implications of automation. Ensuring a fair and equitable transition to a more automated workforce involves careful consideration of social values and the allocation of advantages and liabilities.

In conclusion, public perceptions and misunderstandings are key components of the sociocultural context in which robotic technologies are implemented. The combination between societal influences, media depictions, and individual experiences determines how robots are regarded. Addressing misunderstandings is not only vital for increasing public acceptance but also for leading ethical discourse and influencing policy choices. By encouraging clear communication, engaging the public, and supporting realistic media depictions, society may negotiate the changing connection between humans and robots with better understanding and cooperation.

10.3 Legal Frameworks for Emotional Robots

Legal frameworks for emotional robots constitute a significant component of the expanding area of robotics and artificial intelligence. As emotional robots become increasingly incorporated into numerous parts of society, including healthcare, education, and companionship, problems emerge about their legal status, responsibilities, and ethical implications. Examining current and new legal frameworks is necessary to solve the difficulties brought by the junction of technology and emotions.

One key part of the legal framework for emotional robots concerns upon their categorization. Robots, especially ones meant to express emotions, frequently blur the borders between simple tools and organisms with a degree of autonomy. The legal system historically categorizes things and entities based on predetermined criteria. As emotional robots get more sophisticated, defining whether they should be viewed as property, autonomous agents, or a distinct category entirely becomes a hard task. For instance, in the field of intellectual property, problems emerge concerning the ownership of emotional expressions created by a robot—does it belong to the manufacturer, the programmer, or the end-user?

Liability considerations provide another key factor within the legal context for emotional robots. In instances where robots are supposed to aid or interact with people, the topic of accountability for any damage or mistakes becomes crucial. Legal frameworks must provide clear standards on who bears accountability — the manufacturer, the programmer, the user, or a combination of these groups. A prominent example is the deployment of emotional robots in hospital settings, where they give companionship or aid to patients with mental health disorders. If an emotional robot were to deliver wrong advise or mistakenly cause injury, defining the person liable for any resultant legal implications becomes critical.

Privacy problems also play a significant part in the legal considerations for emotional robots. These robots typically gather and use sensitive personal data to modify their emotional reactions or enhance user experience. Legal frameworks need to address the extent of data collection, storage, and sharing, ensuring that users' emotional data is managed appropriately and ethically. As an instance, emotional robots employed in therapeutic contexts may acquire detailed facts about users' emotional states. Establishing rigorous privacy policies may prevent the exploitation of such information and safeguard consumers from unwanted access or data breaches.

In the context of employment, legal structures must adjust to the growing integration of emotional robots in the labor. The employment of robots in occupations that include emotional labor, such as customer service or care professions, raises problems concerning workers' rights and the possible displacement of human workers. Some governments have started researching laws to guarantee that robots complement human work without generating widespread unemployment. Striking a balance between technological innovation and preserving human employment rights needs legislative frameworks that foresee and manage the socio-economic repercussions of emotional robots in the workplace.

Ethical issues linked with legal frameworks for emotional robots add another dimension of complication. The ethical issues of constructing robots that replicate or elicit human emotions need a considered approach in legal design. For example, restrictions may need to be set to avoid the production of emotional robots that may be abused for nefarious reasons or used to manipulate susceptible humans. Legal frameworks should not only assess the immediate use of emotional robots but also foresee possible misuse and contain protections to avoid damage.

International cooperation becomes vital in creating legal frameworks for emotional robots. Given the global nature of technology development and deployment, harmonizing legislation across countries is necessary to maintain uniformity and efficacy. The issue comes in harmonizing varied cultural, legal, and ethical opinions on the function of emotional robots in society. International organizations and agreements may encourage the transmission of information and best practices, establishing a common awareness of the legal consequences of emotional robots.

As emotional robots grow increasingly ubiquitous, legal frameworks must remain dynamic and adaptive. Anticipating future advances in technology is intrinsically tough, but establishing frameworks that can handle increasing capabilities and applications is vital. The legal system must be ready to face unanticipated ethical challenges and technical breakthroughs, ensuring that emotional robots continue to serve humans in a way that complies with social norms and legal requirements.

In conclusion, the legal frameworks for emotional robots are at the convergence of technology, ethics, and society norms. Addressing categorization, responsibility, privacy, employment, and ethics within legal systems is crucial to traverse the complicated world of emotional robots. The advancement of these frameworks demands a collaborative, multinational effort to reconcile innovation with accountability, encouraging the appropriate development and deployment of emotional robots in a fast evolving technological age.

Chapter 11: Humanoid Robots in Art and Entertainment

11.1 Robots in Film: From Metropolis to Ex Machina

Robots have long been an intriguing topic in the field of film, changing and reflecting society ideas towards technology, artificial intelligence, and the human experience. From the classic "Metropolis" to the more current "Ex Machina," the representation of robots in movies has changed, revealing important insights into the ever-evolving interaction between humans and technology.

"Metropolis," directed by Fritz Lang in 1927, regarded as a pioneering work in the science fiction genre and a classic in cinematic history. The film addresses issues of industrialization, class strife, and the implications of unrestrained technological growth. In "Metropolis," the famous Maria robot is a symbol of both intrigue and horror. The robot adopts the look of the film's heroine, Maria, but with a chilly, mechanical disposition. This early movie depiction lay the groundwork for the portrayal of robots as both wonders of human ingenuity and possible dangers to civilization.

As cinematic technology evolved, so did the representation of robots. In Ridley Scott's "Blade Runner" (1982), based on Philip K. Dick's book "Do Androids Dream of Electric Sheep?," the notion of androids or "replicants" became fundamental to the story. The video tackles concerns regarding the nature of consciousness, morality, and what it means to be human. The replicants, although being artificial, have sophisticated emotions and wants. This change from the merely mechanical robots of prior films signals a turn towards studying the emotional and existential elements of artificial creatures.

The history of robot depictions in movies made another stride with Steven Spielberg's "A.I. Artificial Intelligence" (2001). The film goes into the notion of a mechanical kid capable of understanding love and longing for acceptance. Spielberg's examination of emotions in artificial creatures opened the door for a more complex and empathic depiction of robots in movies. David, the mechanical kid, becomes a mirror reflecting the human need for connection, blurring the distinctions between what is artificial and what is true.

Moving into the 21st century, "Ex Machina" (2014), directed by Alex Garland, delivers a thought-provoking perspective on the interaction between humans and machines. The story focuses on the development of Ava, an artificial intelligence housed in a highly sophisticated humanoid robot. Ava boasts not simply physical strength but also a deep knowledge of human emotions and manipulation. "Ex Machina" pushes the spectator to confront the ethics of creating intelligent entities capable of duplicity and self-preservation. The film posits that the actual threat rests not in the robots themselves but in the intentions and acts of their human creators.

A recurring thread across these films is the examination of the "uncanny valley," a notion that denotes discomfort or anxiety when a robot or humanoid becomes nearly, but not quite, indistinguishable from a human. This notion is heavily depicted in "Metropolis" with the Maria robot and is reflected in following films like "Blade Runner" and "Ex Machina." The issue of producing robots that are both realistic and socially acceptable becomes a recurrent topic, mirroring real-world worries about the integration of artificial intelligence into society.

The representation of robots in movies is not just a reflection of technology progress but also a commentary on society concerns and hopes. The dread of losing control over machines, the ethical implications of creating intelligent creatures, and the philosophical problems surrounding artificial life are all incorporated in these cinematic scenarios. "Metropolis" established the groundwork by presenting the notion of robots as instruments of both creation and destruction, and following films have elaborated on this topic, examining the emotional and ethical elements of artificial intelligence.

In conclusion, the path of robots in movies, from "Metropolis" to "Ex Machina," parallels the growing interaction between people and technology. These cinematic portrayals serve as a prism through which society explores its own attitudes, anxieties, and ethical issues around artificial intelligence. The journey from mechanical creatures to emotionally complex entities in cinema reflects not just the breakthroughs in technology but also the deeper issues of what it means to be human in a society increasingly linked with artificial intelligence. As technology continues to evolve, the cinematic investigation of robots will undoubtedly remain a dynamic and growing representation of our collective views and fears about the incorporation of machines into the fabric of human life.

11.2 Emotional Robotics in Literature and Pop Culture

Emotional robotics has beyond the bounds of labs and technical areas, leaving its imprint in numerous parts of human society, including literature and pop culture. In literature, the integration of emotional robotics has given writers with a new canvas to investigate the junction of human emotions and artificial intelligence. One noteworthy example is Isaac Asimov's "I, Robot," a collection of science fiction novels that looks with the ethical issues of developing emotionally aware robots. Asimov's work provided the framework for debates on the emotional capacity of machines and the possible repercussions of blurring the borders between the human and the robotic.

Pop culture, with its enormous reach and effect, has also embraced the topic of emotional robotics. Films like "Blade Runner" and "Ex Machina" have become classic portrayals of the complications surrounding humanoid robots and their emotional experiences. In "Blade Runner," the figure of Rachael raises issues about the validity of feelings in replicants, generating arguments on the nature of consciousness and the ethical implications of creating creatures with emotional capacity. Similarly, "Ex Machina" tackles the inner torment of an artificial intelligence being called Ava, challenging the borders of empathy and the ramifications of creating computers with the capacity to have emotions.

Literary and cinematic works frequently act as mirrors to social issues and comments on developing technology. They enable audiences to face and explore the possible implications of emotional robots on our collective awareness. The topic has been repeated in numerous genres, from classic science fiction to modern speculative fiction, offering a rich tapestry for investigating the emotional lives of robots and the ramifications of integrating them into human society.

Beyond the written word and cinema, emotional robotics has found a home in various types of popular media, including music and television. Musicians and composers have employed electronic sounds and synthesizers to produce a feeling of emotion, blurring the distinctions between human and machine-generated emotions. In the television series "Westworld," the narrative carefully weaves the emotional experiences of humanoid hosts, causing viewers to question the validity of these feelings and the ethical implications surrounding the usage of artificial humans for entertainment.

Moreover, the subject of emotional robotics has entered the domains of gaming, where interactive narrative lets players to connect with emotionally nuanced characters. Games like "Detroit: Become Human" show a future where androids battle with their newfound feelings, bringing profound concerns about the moral duties related to the development of emotionally aware robots. The participatory aspect of gaming gives a unique platform for players to immerse themselves in storylines that explore the emotional landscapes of artificial beings, building a stronger connection between the audience and the issues conveyed.

Literature and pop culture serve as potent vehicles for conveying ideas and altering public opinions. Through the research of emotional robotics in various cultural spheres, producers have the chance to not only amuse but also prompt meaningful conversations about the ramifications of evolving technology. The representation of emotionally competent robots in literature and pop culture acts as a symbolic prism through which society might analyze its relationship with technology, addressing concerns, and picturing possible futures.

As society grapples with the ethical implications of integrating emotional robots into different facets of everyday life, the narratives portrayed in literature and pop culture become vital in molding public conversation. These innovative works question established preconceptions about the powers and limits of robots, encouraging viewers to face their own prejudices and beliefs. The emotional complexity assigned to robots in these novels compels viewers and readers to consider on what it means to feel emotions and the moral duties attached to developing beings that can replicate such experiences.

In conclusion, the examination of emotional robotics in literature and pop culture has become a dynamic and vital component of the greater discourse about the junction of humans and technology. From classic science fiction books to blockbuster films and interactive games, the depiction of emotionally competent robots serves as a multidimensional mirror of social aspirations, anxieties, and ethical issues. As these cultural representations continue to grow, they serve a significant role in molding views, encouraging discourse, and influencing the continuous development of emotional robotics in the real world.

11.3 The Role of Robots in Performing Arts

The use of robots in the performing arts has risen substantially in recent years, ushering in a new era when technology crosses with creativity on stage. This integration not only challenges conventional concepts of performance but also brings up fresh opportunities for narrative, expression, and audience interaction. Robots have transcended their industrial roots to become active collaborators, contributing to a wide range of creative undertakings.

One prominent area of inquiry is within dance performances. Robots, equipped with advanced sensors and programmed motions, have joined human dancers on stage, creating a captivating interaction between man and machine. A remarkable example is the partnership between acclaimed choreographer Wayne McGregor and Random International, a collection of artists and technologists. Their performance, " +/- Human," featured a robotic being interacting with McGregor's dancers, blurring the lines between human and mechanical movement. This mix of organic and mechanical movements not only captivates viewers but also challenges notions about the boundaries of physical expression.

Similarly, in theatrical plays, robots have usurped roles that were formerly solely held for human performers. The play "Spillikin" by Pipeline Theatre is a striking example. The plot concentrates around the connection between a lady and a robot made by her husband to keep her company after his death. This investigation of love, grief, and the essence of humanity via a human-robot relationship not only serves as a thought-provoking piece of entertainment but also dives into the emotional complexity that develop when artificial beings enter the domain of human experience.

Robots have also found a place in collaborative musical performances. The "Robot Orchestra," designed by Moritz Simon Geist, is an orchestra of robots performing different traditional and electronic instruments. Geist's innovation pushes the traditional bounds of musical composition and performance, illustrating how robots may add distinctive sounds and rhythms to the aural universe. This new method to music making not only demonstrates technical competence but also increases the possibilities of sound expression.

Beyond the stage, robots have delved into immersive and interactive installations, engaging spectators in innovative ways. The Robot Theater Company, established by Japanese roboticist Hiroshi Ishiguro, involves humanoid robots that interact with each other and with the audience. In these works, the border between spectator and performer blurs, as the robots react to human presence and interaction. Such experiences underscore the expanding importance of robots in altering the interaction between art and viewer, ushering in a new age of interactive and responsive creative encounters.

The inclusion of robots in the performing arts also extends to avant-garde projects that push the limits of what is considered conventional performance. "Robot-R-Human" by the Swiss group Companie 111 is a multimodal performance where a robot and a human dancer engage in a compelling discussion. Through this research, the designers question the usual conceptions of story structure, character development, and the spatial dynamics of performance. The combination of robotics with avant-garde aesthetics displays a dedication to pushing creative limits and investigating the unexplored frontiers of human-robot cooperation.

Furthermore, the employment of robots in performing arts has not been confined to large-scale shows. Small-scale performances at experimental theaters and art galleries have also welcomed the integration of robots. These intimate settings give a platform for artists to investigate the complexities of human-robot interaction in greater depth, allowing for a closer investigation of the emotional and psychological effect on the audience.

In addition to the creative world, educational programs have harnessed robots as instruments for teaching and learning within the performing arts. The usage of educational robots, such as NAO and Cozmo, enables students to learn programming, design, and creative expression. By incorporating robots into performing arts education, students get a comprehensive grasp of technology's role in creative expression, producing a new generation of artists who can fluidly negotiate the junction of art and technology.

While the incorporation of robots in the performing arts has brought about revolutionary experiences, it has also spurred questions about the ethical implications and social reflections contained in these creations. As robots become increasingly incorporated into the fabric of creative expression, issues emerge concerning the nature of creativity, the authenticity of emotional expression, and the possible influence on human employment within the arts. These talks contribute to a wider discourse regarding the developing interaction between humankind and technology in the context of creative creativity.

In conclusion, the function of robots in the performing arts has grown into a multidimensional investigation of innovation, expression, and cooperation. From dance performances that blur the barriers between human and machine movement to theater plays that dig into the emotional intricacies of human-robot interactions, robots have become vital contributions to the cultural landscape. As technology continues to evolve, the performing arts stand at the vanguard of innovation, testing preconceptions, and increasing the possibilities of what may be done via the symbiotic partnership of people and robots on stage.

Chapter 12: The Emotional Turing Test

12.1 Defining Success in Emotional Robotics

Defining success in emotional robotics is a multidimensional task that entails analyzing the capacity of robots to accurately copy and react to human emotions. Success, in this context, is not exclusively assessed by the technical correctness of facial expressions or the precision of emotion identification algorithms. Rather, it incorporates the larger influence of emotional robots on human experience, their power to facilitate communication, and their potential to contribute positively to numerous sectors.

One essential part of success in emotional robotics is in the congruence between the desired emotional expression of the robot and the human interpretation of such emotions. A good emotional robot should express emotions in a way that is not only recognized but also connects with human viewers. For instance, if a robot is created to exhibit empathy in caregiving circumstances, success would be attained if its emotions inspire a real emotional reaction from people, establishing a feeling of connection and understanding.

The efficacy of emotional robots is partly dependant on their flexibility to varied cultural and individual variances in emotional expressiveness. Success in emotional robotics includes designing systems that can negotiate the many intricacies of cultural variances in the perception of emotions. For example, a robot deployed in healthcare settings across various cultures should be capable of altering its expressions to correspond with the varying expectations and conventions associated to emotional communication.

Furthermore, success in emotional robots goes beyond the technology arena to incorporate ethical problems. An emotionally intelligent robot must be created and constructed with ethical rules that value the well-being and privacy of people. For instance, a robot aiding persons with mental health difficulties should be trained to handle sensitive information responsibly and adhere to rigorous confidentiality requirements, therefore assuring the ethical use of emotional data.

Real-world application is a fundamental prerequisite for success in emotional robotics. The influence of emotional robots may be witnessed in numerous fields, such as healthcare, education, and customer service. In healthcare, for example, a robot accompanying patients during rehabilitation sessions would be judged successful if it successfully encourages and engages people, leading to better overall well-being.

Moreover, the success of emotional robots is inextricably tied to its integration into human civilization. A successful emotional robot is one that smoothly integrates into the everyday lives of humans, boosting human-machine cooperation. For instance, a companion robot built for older folks should not only demonstrate emotional intelligence but also integrate into the social fabric of the society, encouraging a sense of camaraderie and eliminating feelings of loneliness.

The function of emotional robots in encouraging pleasant human-machine interactions is fundamental to their success. In customer service, for instance, a robot taught to comprehend and react correctly to human emotions may boost user pleasure. Success in this context is not only assessed by the robot's ability to address issues but also by its power to express empathy and generate a pleasant customer experience.

Moreover, progress in emotional robotics is intimately connected to breakthroughs in artificial intelligence and machine learning. The constant refinement of algorithms that allow robots to learn and change their emotional reactions based on experience is vital to their success. For example, a robot developed for educational purposes should be capable of dynamically altering its teaching method depending on the emotional reactions and learning patterns of individual pupils, leading to more effective and individualized instruction.

The idea of success in emotional robotics also entails the construction of benchmarks and criteria for measuring the emotional intelligence of robots. Research and commercial initiatives to define and evaluate emotional intelligence in robots add to the overall success of this sector. A unified framework for analyzing emotional robotics guarantees that gains are achieved in a logical and quantitative way, allowing the comparison and development of different emotional robots across varied applications.

Success in emotional robotics is not static; it changes with the dynamic nature of human-robot interactions. Continuous user input, iterative design methods, and the capacity to learn from both triumphs and mistakes contribute to the continuous development of emotional robots. For instance, a robot deployed in educational settings may undergo ongoing upgrades based on input from instructors and students, enabling it to adapt and develop over time.

In conclusion, defining success in emotional robotics exceeds technical precision and enters into the domain of human experience, cultural adaptation, ethical issues, and real-world application. The success of emotional robots is inextricably tied to their capacity to effectively transmit and react to human emotions, facilitating meaningful interactions across varied situations. As emotional robotics continues to evolve, the pursuit of success in this sector demands a comprehensive strategy that merges technical innovation with a thorough knowledge of human emotions and social requirements.

12.2 Evaluating Emotional Intelligence in Robots

Emotional intelligence in robots has been a key area in the subject of human-robot interaction, as researchers strive to construct computers capable of understanding and reacting to human emotions. This assessment procedure entails examining a robot's capacity to identify, comprehend, and correctly react to emotional stimuli, mimicking the multifaceted nature of human emotional intelligence. Achieving this needs a complex strategy, including many variables such as facial expression detection, tone of voice analysis, and adaptive behavior. The efficacy of emotional intelligence in robots is vital for their smooth integration into human surroundings, where social and emotional interactions play key roles.

One key part of measuring emotional intelligence in robots is the identification of facial emotions. Mimicking the human capacity to perceive emotions via facial signals, researchers have built algorithms and sensors that allow robots to evaluate and react to human facial expressions effectively. For instance, a robot equipped with powerful face recognition technology can discern between a grin signaling satisfaction and a frown suggesting disapproval. This functionality helps the robot to react correctly, promoting a more natural and intuitive contact with people.

Tone of voice analysis is another crucial component in judging emotional intelligence in robots. Just as humans identify emotional subtleties via vocal intonations, robots with superior speech processing skills may evaluate the tone, pitch, and rhythm of human speech. For example, a robot built to perceive grief in a person's voice may change its replies with empathy, indicating a heightened degree of emotional intelligence. This capacity to perceive emotional nuances in speech helps considerably to the robot's performance in social interactions.

Adaptive behavior is a significant consequence of emotional intelligence assessment in robots. An emotionally intelligent robot should not only sense emotions but also adjust its behavior appropriately. Through machine learning algorithms and artificial intelligence, robots may learn from human interactions and adjust their reactions over time. For instance, a social companion robot may learn to offer consolation to a person who exhibits melancholy while altering its conversational style dependent on the user's emotional state. This flexibility boosts the robot's potential to participate in emotionally resonant and culturally relevant interactions.

To further boost emotional intelligence in robots, researchers are researching the integration of affective computing. Affective computing includes giving robots with the capacity to sense and react to human emotions, going beyond typical programming paradigms. Machine learning algorithms allow robots to continually learn from their interactions, boosting their emotional awareness and reactivity. This dynamic learning process is comparable to human emotional growth, where experiences affect an individual's emotional intelligence over time.

Despite these developments, obstacles exist in judging emotional intelligence in robots. One noteworthy problem is the cultural and individual variety in human emotions. Different cultures may express and perceive emotions in distinct ways, requiring robots to demonstrate a degree of cultural sensitivity. Additionally, humans differ in their emotional displays, making it challenging for robots to generalize emotional signals appropriately. Overcoming these issues needs significant data collecting and the ongoing refining of algorithms to handle varied emotional responses and cultural variations.

Ethical issues are crucial to the assessment of emotional intelligence in robots. As these computers grow increasingly interwoven into human life, problems emerge regarding privacy, permission, and the possible manipulation of emotions. Ensuring that robots follow ethical limits in emotional interactions is vital to creating trust between people and technology. Striking the correct balance between emotional response and ethical concerns is an ongoing subject of study and development in the field of emotional intelligence in robots.

The real-world applications of emotionally intelligent robots are numerous and significant. In healthcare settings, robots with emotional intelligence may give companionship and assistance to persons facing loneliness or emotional discomfort. Educational robots equipped with emotional awareness may alter their teaching approaches to respond to the emotional requirements of pupils, producing a more customized and effective learning experience. Furthermore, in customer service and care businesses, emotionally intelligent robots may boost the quality of interactions, delivering compassionate solutions to consumer inquiries or concerns.

In conclusion, the assessment of emotional intelligence in robots is a complex task that covers facial expression detection, tone of voice analysis, adaptive behavior, and cultural sensitivity. Advancements in machine learning, affective computing, and artificial intelligence lead to the creation of robots that can comprehend and react to human emotions with increasing complexity. However, problems such as cultural heterogeneity and ethical concerns underline the need for continued study and development in this emerging subject. The real-world applications of emotionally intelligent robots offer the possibility of changing different sectors, from healthcare and education to customer service, ushering in an era when computers can connect with people in emotionally resonant and socially savvy ways.

12.3 Challenges in Developing a Standardized Emotional Turing Test

Developing a standardized Emotional Turing Test involves various issues in the domain of artificial intelligence and human-robot interaction. The complexity of human emotions, the subjective nature of emotional experiences, and the inherent cultural heterogeneity in expressing and interpreting emotions all add to the complication of developing a globally acknowledged baseline for judging emotional intelligence in computers. One of the greatest obstacles comes in creating a comprehensive set of criteria that may successfully capture the complexities of emotional comprehension and expression.

One key difficulty in designing a standardized Emotional Turing Test is the variable and context-dependent nature of human emotions. Unlike conventional Turing Tests that concentrate on objective measurements of intelligence, emotions are fundamentally subjective and impacted by different variables such as personal experiences, cultural background, and individual variances. For instance, a robot could properly detect and react to emotions of pleasure in one cultural context but fail to comprehend the same feeling in a different cultural environment where expressive conventions vary. The difficulty, however, is to create a test that acknowledges this inherent subjectivity while yet delivering a valid rating of emotional intelligence.

Closely connected to the subjectivity of emotions is the difficulty of generating a uniform collection of emotional stimuli. Emotions are multidimensional and may emerge in different ways, including facial expressions, body language, tone of voice, and even physiological changes. Designing a test that captures this variation while being consistent across diverse events and cultural settings is a hard undertaking. For example, a robot could excel at identifying facial expressions linked with delight but fail to read subtle verbal signals indicative of the same emotion. Striking a balance between comprehensiveness and uniformity is important for the success of an Emotional Turing Test.

Moreover, the difficulty extends to the dynamic character of emotions. Unlike static IQ ratings, emotions develop throughout time and may be impacted by changing circumstances. A robot capable of reliably recognizing an emotion in one instant may confront difficulty in responding to variations in emotional states or grasping the evolution of emotional experiences. For instance, identifying the shift from surprise to disappointment or distinguishing the minor distinctions between comparable emotions like concern and worry offers a problem that needs a degree of temporal knowledge beyond the capability of many contemporary AI systems.

Another challenge in designing a standardized Emotional Turing Test is the ethical issue linked with emotional manipulation. As robots grow increasingly competent at identifying and reacting to human emotions, there emerges the possibility for purposeful or unintended manipulation of emotions for diverse objectives. For example, a robot created to give emotional support may need to strike a fine balance between delivering true empathy and avoiding manipulation. Crafting a test that analyzes not just the accuracy of emotional detection but also the ethical implications of emotional responses is vital for responsible AI development.

Cultural biases further aggravate the problems of designing a standardized Emotional Turing Test. Emotions are expressed and understood differently across cultures, creating a degree of complexity that deserves careful study. A robot trained on a dataset from one cultural context may fail to appropriately recognize and react to emotions in persons from a different cultural background. Ensuring cross-cultural validity in an Emotional Turing Test involves a rigorous approach to dataset curation and algorithm training, identifying and reducing biases to produce fair and impartial judgments of emotional intelligence.

Moreover, the difficulty extends to the dynamic character of emotions. Unlike static IQ ratings, emotions develop throughout time and may be impacted by changing circumstances. A robot capable of reliably recognizing an emotion in one instant may confront difficulty in responding to variations in emotional states or grasping the evolution of emotional experiences. For instance, identifying the shift from surprise to disappointment or distinguishing the minor distinctions between comparable emotions like concern and worry offers a problem that needs a degree of temporal knowledge beyond the capability of many contemporary AI systems.

As technology improves, the difficulty of explainability becomes more crucial in the context of emotional AI. Understanding how an AI system arrives to a certain emotional conclusion is vital not just for transparency but also for addressing possible biases and inaccuracies. Developing a standardized Emotional Turing Test entails addressing the interpretability of algorithms to guarantee that the decision-making processes are not opaque. This difficulty is especially prominent in emotionally charged settings when the repercussions of misunderstanding might be considerable.

In conclusion, the obstacles in designing a standardized Emotional Turing Test are profoundly based in the subjective, dynamic, and culturally affected character of human emotions. From setting thorough criteria to negotiating the ethical consequences of emotional manipulation, each roadblock emphasizes the complexities of judging emotional intelligence in robots. As technology improves, resolving these difficulties becomes crucial for the ethical development of emotionally intelligent AI systems, opening the path for a future where robots can truly comprehend and react to human emotions across varied situations.

156

Chapter 13: Interdisciplinary Approaches to Emotional Robotics

13.1 Collaboration Between Robotics and Psychology

Collaboration between robotics and psychology is a dynamic junction of two different sciences, driving improvements in both technology and human understanding. This synergy attempts to better the development and deployment of robots by integrating psychological concepts, and conversely, to use robotic technology for psychological research and applications.

The partnership is obvious in the inclusion of psychological ideas into the design of robots, especially those meant for social interaction. For instance, robots developed to aid persons with autism spectrum disorders sometimes depend on ideas from social psychology. These robots are built to understand and react to social signals, helping users enhance their social communication abilities. By incorporating psychological insights into the programming of these robots, designers aspire to develop tools that engage with human users on a deeper emotional and cognitive level.

Furthermore, psychology contributes to the creation of robots capable of comprehending and reacting to human emotions. Emotion detection is a fundamental part of human-robot interaction, and psychology study on facial expressions, body language, and verbal signals influences the algorithms inherent in these robots. For example, social robots utilized in healthcare contexts are meant to understand and react properly to the emotional states of patients, offering empathic treatment. This partnership between robotics and psychology underscores the necessity of understanding human emotions for the efficient integration of robots into numerous social areas.

In addition to the influence on technology, the relationship is crucial in harnessing robots for psychological study. Robotics offers psychologists with unique instruments to perform tests and investigations that were previously impossible. Humanoid robots, for instance, may be built to imitate certain social circumstances, enabling researchers to watch and evaluate human behavior in controlled conditions. This promotes a greater knowledge of social dynamics, decision-making processes, and the psychological aspects impacting human behavior.

Moreover, the usage of robots in therapeutic settings emphasizes the collaboration's potential in tackling mental health concerns. Psychologists are studying the incorporation of robotic companions in therapeutic therapies, especially for persons feeling loneliness or emotional discomfort. Paro, a robotic seal intended for therapeutic reasons, shows this teamwork. Its realistic look and responsive activity intended to inspire emotional reactions, creating a feeling of warmth and friendship. Such applications highlight how robots may support psychological therapy, delivering creative ways to meet social issues.

The partnership extends to the examination of ethical concerns surrounding human-robot interaction. Psychologists play a significant role in investigating the psychological effect of robotics on people and society. Understanding how people perceive and react to robots, as well as the ethical implications of incorporating robots into everyday life, needs a complex psychological approach. For instance, research addressing the idea of the "uncanny valley" look into the uneasiness individuals may experience when robots closely resemble humans but lack key realistic traits. This psychological concept influences robotic design ideas to promote user acceptance and good interactions.

Furthermore, the partnership between robots and psychology is visible in the area of cognitive psychology, where researchers employ robotic systems to examine cognitive processes and memory. Humanoid robots outfitted with artificial intelligence are deployed as experimental instruments to examine how people perceive, process, and recall information. This multidisciplinary method gives a unique viewpoint on cognition, enabling psychologists to explore human cognitive processes via the prism of robotic activity.

The educational area also benefits from the partnership between robots and psychology. Robots are being employed in educational settings to boost learning experiences and engagement. Psychological theories of learning and motivation underlie the creation of educational robots meant to adapt to individual learning styles and deliver individualized feedback. The incorporation of psychological concepts into educational robots guarantees that these technologies conform with recognized pedagogical procedures and contribute favorably to the learning outcomes of students.

Moreover, the cooperation is represented in the area of human-robot teaming, where psychologists and roboticists work together to maximize collaboration between people and robots. Understanding human elements, such as trust, communication, and decision-making, is vital for establishing successful human-robot collaborations. Psychologists give insights on team dynamics, communication patterns, and the influence of automation on human performance. This multidisciplinary cooperation strives to build synergistic collaborations that use the skills of both people and robots in varied work situations.

In conclusion, the partnership between robotics and psychology constitutes a dynamic connection with important consequences for technology, research, and social well-being. As robots become more incorporated into many facets of human life, understanding the psychological roots of human-robot interaction becomes crucial. This partnership not only helps the creation of more socially competent and emotionally intelligent robots but also gives psychologists with unique tools to investigate and stretch the limits of human cognition and behavior. The continuous cooperation between these domains offers the prospect of building a future where robots and humans live together, informed by a profound grasp of both technical and psychological elements.

13.2 Integrating Neuroscience in Emotional AI

The integration of neuroscience into the sphere of Emotional AI marks a substantial confluence of two independent but related sciences. This merger not only enriches our knowledge of human emotions but also promotes the creation and use of artificial intelligence with emotional intelligence capabilities.

At the center of this integration lies the quest to mimic and comprehend the complicated systems of the human brain responsible for emotional processing. Neuroscientific findings into the brain underpinnings of emotions, such as the involvement of the limbic system and the amygdala, have offered a basic knowledge. Researchers want to transfer these neurological processes into algorithms and computer models that may replicate emotional reactions in AI systems.

One important component of incorporating neuroscience into Emotional AI is the effort to imitate the human brain's capacity to sense and analyze facial emotions. For instance, research in neuroscience have revealed certain areas, such as the fusiform face area, that are devoted to facial identification. By combining these discoveries, Emotional AI systems may be created to perceive and react to a varied variety of facial expressions, allowing more complex and contextually appropriate emotional interactions.

Furthermore, neuroscientific research on the brain underpinnings of empathy has inspired the creation of AI systems capable of identifying and reacting empathetically to human emotions. For instance, mirror neurons, which are involved in the capacity to perceive and share the emotions of others, serve as a model for constructing AI systems that can replicate sympathetic reactions. This integration has potential for applications in healthcare, customer service, and numerous human-centric sectors where sympathetic engagement is vital.

The notion of brain-computer interfaces (BCIs) is another area where neuroscience overlaps with Emotional AI. BCIs, by directly linking the human brain to external equipment, present a unique route for analyzing and extracting emotional data. Integrating these signals into Emotional AI systems offers up opportunities for more tailored and adaptable emotional reactions. For example, an AI-driven virtual assistant may dynamically modify its tone and replies depending on real-time emotional data received via a BCI, enabling a more customized and human-like engagement.

The discipline of affective computing, which focuses on imbuing computers with the capacity to comprehend and react to human emotions, mainly depends on findings from neuroscience. Emotion identification algorithms, a major component of affective computing, generally take inspiration from neuroscientific research that uncover patterns of brain activity linked with certain emotions. This integration enables Emotional AI systems to not only detect fundamental emotions but also grasp more complicated emotional states, such as ambiguity or mixed feelings, mimicking the intricate nature of human emotional experiences.

Moreover, the integration of neuroscience in Emotional AI has ramifications for mental health applications. Understanding the neurological basis of stress, anxiety, and other mental states allows the creation of AI-driven solutions for early diagnosis and intervention. For instance, an Emotional AI system informed by neuroscientific research could analyze subtle changes in speech patterns or facial expressions to identify signs of emotional distress, providing timely support or intervention.

However, the integration of neuroscience into Emotional AI is not without its hurdles. The complexity of the human brain, with its extensive neural networks and subjective nature of emotional experiences, offers difficulties for constructing accurate and universally applicable models. Additionally, ethical issues regarding privacy and the proper use of neuroscientific data in AI applications must be carefully managed.

In conclusion, the integration of neuroscience in Emotional AI reflects a symbiotic connection between two disciplines that, when united, have the potential to change human-machine interactions. From facial expression detection to sympathetic reactions and individualized interactions, the insights garnered from neuroscience offer the building blocks for increasingly complex and emotionally intelligent AI systems. As technology continues to improve, the cooperation between neuroscience and Emotional AI offers the possibility of not just advancing our knowledge of human emotions but also building AI systems that truly connect with and react to the subtleties of the human emotional experience.

13.3 The Role of Design Thinking in Emotional Robotics

The importance of design thinking in emotional robotics is crucial, involving a multidisciplinary approach that goes beyond conventional utility. Design thinking promotes empathy, creativity, and iterative problem-solving to handle complex difficulties, fitting flawlessly with the intricate nature of building robots with emotional capacities. In this scenario, empathy is not just oriented towards end-users but also extends to comprehending the social and cultural environments in which these robots will work.

Design thinking fosters a user-centric approach, ensuring that emotional robots are not merely technologically sophisticated but also connect with human emotions and preferences. One significant example is the creation of socially assistive robots meant to give companionship and assistance for persons with special needs. By utilizing design thinking methods, engineers and designers may obtain insights into the emotional needs of users, developing a greater knowledge of how these robots might positively influence the lives of humans.

The iterative nature of design thinking is especially advantageous in the domain of emotional robots. Engineers may build and test numerous versions of a robot's emotional expressions, adjusting them depending on user input. This iterative technique allows for continual refinement, correcting any differences between the intended emotional signals and the users' impressions. Through this constant refining, designers may strengthen the robot's emotional intelligence, making it more receptive to the various and dynamic nature of human emotions.

An exemplary example of design thinking in emotional robotics is the construction of robots for therapeutic reasons, such as treating persons with autism spectrum disorder. Designers utilize empathy to understand the particular emotional requirements of persons with autism, adjusting the robot's emotions and interactions to give a helpful and pleasant experience. Through prototyping and user testing, designers may optimize the robot's actions, ensuring that it fosters favorable emotional responses in users.

Furthermore, design thinking plays a significant role in resolving ethical problems in emotional robots. As these robots grow increasingly interwoven into all facets of human life, ethical problems involving privacy, permission, and the possible manipulation of emotions come to the forefront. Design thinking fosters a proactive approach to ethics, urging designers to examine these ethical issues from the beginning phases of creation. For instance, designers may integrate features that emphasize user privacy or provide explicit standards on emotional manipulation to enable responsible and ethical usage of emotional robots.

Cultural sensitivity is another facet where design thinking excels in the domain of emotional robots. Different cultures may perceive and express emotions in unique ways, and a one-size-fits-all approach may lead to misunderstandings or discomfort. Design thinking enables designers to participate in cross-cultural study, learning how emotional expressions are received in varied cultural situations. This cultural understanding is vital in the creation of emotionally intelligent robots, ensuring that their expressions correspond with societal norms and preferences.

In the education industry, emotional robots are being built to aid students in learning situations. Design thinking enables for the customisation of these robots to respond to the emotional and intellectual requirements of varied groups of youngsters. By integrating educators, child psychologists, and other stakeholders in the design phase, emotional robots may be tuned to give optimum assistance for diverse learning styles and emotional sensitivities.

Moreover, design thinking facilitates cooperation among people with various experience. Engineers, psychologists, ethicists, and designers may work cooperatively, utilizing their distinct views to build emotionally intelligent robots that not only operate effortlessly but also conform to ethical norms and engage with people on an emotional level. This collaborative approach is shown in initiatives where multidisciplinary teams get together to build robots that aid in emotional therapy or companionship for the elderly.

In conclusion, the function of design thinking in emotional robots is multidimensional, embracing empathy, iterative prototyping, ethical concerns, cultural sensitivity, and multidisciplinary cooperation. Through a design thinking perspective, engineers and designers may construct emotionally aware robots that not only satisfy functional needs but also connect with users in varied cultural and social situations. As emotional robotics continues to advance, the use of design thinking concepts will remain crucial in navigating the complicated environment of human-robot interaction.

Chapter 14: Public Perception and Acceptance of Emotional Robots

14.1 Shaping Positive Attitudes Towards Emotional Robots

Shaping favorable views towards emotional robots entails a complex strategy that covers psychological, sociological, and ethical factors. The perception of robots, especially those meant to express emotions, is impacted by several elements, including cultural origins, personal experiences, and the amount of information about the capabilities and limits of these machines.

At the psychological level, creating optimistic attitudes entails tackling the phenomena known as the "uncanny valley." This notion, proposed by roboticist Masahiro Mori, posits that as a robot's look gets more human-like, there is a threshold at which it evokes discomfort or disquiet in human viewers. To counteract this, designers must carefully negotiate the design space, achieving a balance that generates a favorable and comfortable reaction. For instance, a research by Bartneck et al. (2007) digs into the difficulty of avoiding the uncanny valley and underlines the necessity of purposeful design decisions that create positive emotional reactions.

Societal views also play a vital influence in developing attitudes towards emotional robots. Cultural variables considerably effect how humans understand and accept robots in their everyday lives. For example, in Japan, where the notion of anthropomorphism is profoundly embedded in culture, there is often higher acceptance of robots showing emotions. In contrast, Western societies may first approach emotional robots with mistrust. Therefore, efforts for fostering positive attitudes need to be culturally sensitive and aware of other viewpoints.

Ethical issues additionally contribute to the forming of views towards emotional robots. Ensuring openness in the design and operation of these robots is crucial. Users need to comprehend the motives behind the emotional expressions of robots and be satisfied that their interactions are founded on ethical norms. For instance, adding ethical criteria in the creation process, as outlined by Picard (2000), may help foster trust and favorably affect views of emotional robots.

To put these concepts in action, consider the case study of PARO, a therapeutic robot meant to mimic a young harp seal. PARO has been deployed in hospital settings to give companionship to patients, especially those with dementia. Research by Dautenhahn and Werry (2004) reveals how PARO's emotional expressions and responsiveness have prompted positive reactions from users, leading to greater emotional well-being. This case demonstrates the potential of emotional robots to positively affect human lives when created with sensitivity and care for user requirements.

Public participation and education are key components of establishing good perceptions about emotional robots. Initiatives that explain the technology and illustrate its advantages may lead to a more educated and tolerant society. Educational programs, seminars, and interactive demonstrations help bridge the information gap and dispel misunderstandings about the capabilities and intentions of emotional robots. By actively including the public in the development process, as recommended by Wood (2007), a feeling of ownership and participation may be developed, leading to more favorable sentiments.

However, problems exist in the form of anxiety and confusion over the integration of emotional robots into everyday life. Science fiction and popular media frequently depict robots as possible dangers, adding to unfavorable sentiments. Addressing these challenges needs a proactive strategy from developers, politicians, and educators. Open discourse and clear communication regarding the ethical issues, safety precautions, and intended uses of emotional robots are crucial to ease worries and develop a foundation of confidence.

Moreover, the integration of emotional robots into numerous areas, such as healthcare and education, might influence public opinions. As emotional robots establish their usefulness in aiding persons with therapeutic needs, boosting education, and increasing general well-being, the narrative around them moves from skepticism to acceptance. Real-world instances of emotional robots making beneficial contributions, combined with honest communication, might eventually normalize their presence in society.

In conclusion, building good views towards emotional robots is a multifaceted process that involves a full knowledge of psychological, sociological, and ethical aspects. By tackling the uncanny valley, being culturally sensitive, complying to ethical rules, and actively engaging and educating the public, developers and researchers may contribute to a more favorable reception of emotional robots. Real-world applications, such as the employment of PARO in healthcare, underline the potential advantages of these robots, underlining the significance of responsible design and public communication in encouraging acceptance and confidence. Through these efforts, emotional robots may move from being innovative technical entities to important companions and contributions to human well-being.

14.2 Addressing Fear and Uncertainty

Addressing anxiety and uncertainty is an important factor in the development and integration of emotional robots into society. As these advanced robots grow more common, recognizing and reducing the apprehensions that humans may carry is vital for encouraging acceptance and cooperation. One big cause of worry is the uncertainty surrounding the capabilities and intentions of these robots.

People typically fear the unfamiliar, and in the case of emotional robots, doubt stems from a lack of familiarity with their design, operation, and ethical constraints. For instance, in the field of healthcare, where emotional robots are increasingly utilized for patient care and companionship, humans may be anxious about the amount of autonomy these robots possess in making vital judgments. To overcome such concerns, developers and authorities must publicly convey the constraints and norms controlling the activities of emotional robots, comforting people and fostering confidence.

Ethical problems are another level of anxiety and uncertainty in the creation of emotional robots. Questions regarding privacy, data security, and possible exploitation of emotional data might cause fear among users. For instance, if emotional robots are deployed in educational settings to aid children with learning issues, parents may concern about the collecting and use of their child's emotional data. Developers need to develop rigorous ethical frameworks, following to tight criteria to preserve user privacy and guarantee appropriate use of emotional data, therefore soothing anxieties and generating a feeling of security.

Cultural variances can add to dread and ambiguity about emotional robots. Different civilizations may have differing viewpoints on human-robot relationships, affected by cultural norms, beliefs, and traditions. For example, a society that strongly values personal privacy and emotional closeness may regard emotional robots differently from a culture that is more receptive of technology interference in personal life. To overcome these issues, developers should participate in cross-cultural research and construct emotional robots that respect and align with varied cultural sensibilities.

The depiction of robots in popular culture and literature further intensifies dread and uncertainty. Sci-fi storylines commonly show robots as dangers to humans, capable of revolt or terrible activities. This representation impacts public perception, leading to fears about the possible risks of emotional robots. Developers and researchers must aggressively interact with the media to give a fair and realistic image of emotional robots, highlighting their positive contributions and the ethical frameworks in place to avoid exploitation.

Moreover, worry and uncertainty may also stem from economic concerns, especially about job relocation. As emotional robots are incorporated into numerous businesses for jobs like as customer service or companionship, there is a persistent fear about job losses. To overcome this, it is vital to stress the collaborative aspect of human-robot collaborations, emphasizing that emotional robots are created to enhance human talents, not replace them. Additionally, measures for retraining and upskilling programs may help relieve worries about job displacement and contribute to a more favorable reception of emotional robots in the workforce.

Education has a crucial role in eliminating anxiety and ambiguity regarding emotional robots. Increasing public understanding about the advantages, limits, and ethical issues involved with these technologies is crucial. Workshops, seminars, and educational campaigns may educate people with the information required to make informed choices and refute stereotypes about emotional robots. By developing a greater grasp of the technology, developers may contribute to a more responsive and educated audience.

In conclusion, tackling anxiety and uncertainty in the context of emotional robots demands a multidisciplinary strategy. Developers must stress openness, ethical concerns, and cultural sensitivity in their ideas and implementations. Engaging with the media to portray a balanced story, addressing economic concerns via education and retraining efforts, and highlighting the collaborative nature of human-robot relationships are essential steps in developing trust and acceptance. Ultimately, the effective integration of emotional robots into society rests on the capacity to manage and soothe the anxieties and uncertainties that unavoidably come with the introduction of disruptive technology.

14.3 Public Engagement in Robotic Emotional Design

Public involvement in robotic emotional design is a vital part that entails incorporating the ideas, worries, and expectations of the public into the creation and implementation of emotional aspects in robots. This method emphasizes the potential social effect of emotional robots and strives to build a collaborative partnership between designers, developers, and end-users. Public participation not only guarantees that robotic emotional design corresponds with ethical norms and cultural values but also promotes acceptance and comprehension of these technologies.

One major part of public interaction in robotic emotional design is the consideration of cultural viewpoints. Different cultures have distinct standards and expectations surrounding emotions, interpersonal interactions, and technology. Public engagement programs entail gathering feedback from varied populations to understand their cultural subtleties and preferences. For example, a robot intended for emotional caregiving in Japan may need to display distinct emotions and actions compared to a comparable robot in the United States owing to cultural variances in emotional expression and social standards.

Moreover, public participation provides as a means for resolving ethical problems linked with robotic emotional design. As emotional robots become increasingly integrated into all sectors of society, ethical problems emerge addressing issues such as privacy, consent, and the possible manipulation of emotions. Engaging the public in debates about these ethical issues helps set norms and standards that emphasize the well-being and rights of persons engaging with emotional robots. For instance, including the public in the establishment of rules for emotional data privacy guarantees that the use of personal emotional information is ethically controlled.

Public interaction also plays a significant role in setting the expectations of end-users and dispelling myths regarding emotional robots. By giving explicit information about the capabilities and limits of these robots, designers may manage user expectations and avert possible disappointment or mistrust. For example, a public engagement campaign can feature interactive seminars or online forums where folks can learn about the technology, ask questions, and obtain a realistic grasp of what emotional robots can and cannot do.

Furthermore, integrating the public in the design process develops a feeling of ownership and empowerment. When people feel that their contribution is acknowledged and included into the creation of emotional robots, they are more inclined to adopt and trust these technologies. This collaborative approach increases the user experience and helps to the effective integration of emotional robots into numerous social sectors. For instance, a project integrating public involvement in the design of companion robots for the elderly might result in characteristics that correspond better with the tastes and requirements of the target users.

Public involvement is not a one-time event but a continuous activity that adapts to the dynamic environment of technology and cultural expectations. As robotic emotional design continues to evolve, engaging the public becomes a dynamic and iterative process that addresses increasing ethical problems, technology breakthroughs, and changing cultural perspectives. For example, when new emotional algorithms are launched, public engagement campaigns might request input on the perceived emotional authenticity of robotic emotions and actions.

However, public interaction in robotic emotional design is not without its problems. Ensuring meaningful and representational involvement from various demographic groups needs careful planning and communication activities. Overcoming possible biases in the selection of participants is vital to prevent accidentally privileging some opinions over others. Additionally, regulating the balance between public input and expert knowledge creates a problem, since technical skill is sometimes essential to grasp the nuances of emotional algorithms and robotic functions.

In conclusion, public interaction in robotic emotional design is a varied and necessary process that contributes to the responsible and ethical creation of emotional robots. By considering cultural views, resolving ethical problems, setting user expectations, and promoting cooperation, designers may build emotional robots that correspond with social norms and promote human-robot interaction. As technology continues to grow, continued public engagement activities will be vital in navigating the complicated terrain of emotions in robotics and ensuring that these technologies contribute positively to the well-being of people and society at large.

Chapter 15: The Ethical Imperative in Emotional Robotics

15.1 Creating Ethical Guidelines for Emotional AI

Creating ethical rules for Emotional AI is a significant task that demands careful evaluation of the possible influence on people and society as a whole. The incorporation of emotions into artificial intelligence systems involves deep ethical challenges, ranging from privacy concerns to the possible manipulation of human emotions. Establishing clear ethical principles is vital to guarantee the proper development and deployment of Emotional AI technology.

One significant ethical concern is the problem of user permission and data privacy. Emotional AI frequently depends on the gathering and analysis of personal data to recognize and react to human emotions properly. Ethical principles must emphasize user permission, ensuring that consumers are fully informed about the data being gathered, how it will be used, and have the ability to opt out if they so choose. For example, a social robot equipped with Emotional AI should get explicit authorization before recording and analyzing the emotional expressions of persons in its surroundings.

Transparency in the design and operation of Emotional AI systems is another key part of ethical norms. Users should have a clear grasp of how these systems operate, including the algorithms utilized to sense emotions and make judgments based on them. Providing transparency not only builds user trust but also enables for outsider monitoring and assessment of any biases in the AI system. For instance, if a face recognition system incorporated in Emotional AI demonstrates prejudice against particular demographic groups, it becomes vital to fix such errors publicly and ethically.

Guarding against the possible abuse of Emotional AI is a key ethical problem. Developers must create standards to prohibit the exploitation of this technology for manipulative or exploitative objectives. For example, Emotional AI should not be deployed to purposely manipulate emotions for commercial advantage or to mislead people. Clear ethical criteria should specify acceptable and unethical applications of Emotional AI, ensuring its deployment conforms with ethical norms and social values.

Bias in Emotional AI systems is a widespread ethical concern that demands careful study. If the training data used to construct these systems is skewed, it might result in unjust and discriminating outputs. Ethical rules should entail extensive testing for biases and the execution of steps to reduce and remedy any found biases. For instance, if an Emotional AI system is largely trained on data from a single demographic group, it may fail to effectively perceive and react to the emotions of persons from underrepresented groups.

An crucial part of ethical principles for Emotional AI entails considering the possible effect on vulnerable groups. For instance, the application of Emotional AI in sensitive environments like as healthcare or education needs careful evaluation of the possible repercussions on vulnerable persons. Ethical standards should underline the need for ethical and sympathetic usage of Emotional AI in such situations, ensuring that the technology increases well-being without compromising privacy or perpetuating current imbalances.

The responsibility of developers and organizations adopting Emotional AI is a cornerstone of ethical principles. Developers should be held responsible for the choices made by their AI systems, and processes for resolving mistakes, biases, or unexpected effects must be in place. An example of this idea in action would be a corporation taking immediate action to address any faults found in its Emotional AI system that may have accidentally caused damage or perpetuated unjust treatment.

Interdisciplinary cooperation is a critical aspect in the formulation of ethical principles for Emotional AI. Involving specialists from many domains, including ethics, psychology, sociology, and law, enables a thorough and well-rounded approach to tackling ethical dilemmas. For instance, engaging with psychologists might give vital insights into the possible psychological effect of Emotional AI on persons, defining ethical rules that emphasize mental well-being.

Global norms and international collaboration are crucial in the formulation of ethical criteria for Emotional AI. As AI technologies transcend geographical borders, universal ethical criteria may promote a unified approach to the responsible development and deployment of Emotional AI. An example is the cooperation of international organizations and governments to construct a framework that maintains uniform ethical standards across various locations, promoting a worldwide culture of responsibility in AI research.

Continuous examination and adaption of ethical principles are necessary to keep pace with the growing world of Emotional AI. As technology evolves and society perceptions alter, ethical rules must remain dynamic and adaptable to developing concerns. Regular changes to guidelines may integrate lessons gained from real-world deployments, ensuring that ethical issues stay pace with technology improvements.

In conclusion, the formulation of ethical standards for Emotional AI is a complicated project that needs a comprehensive approach to meet the various ethical concerns related with this technology. From assuring user permission and transparency to minimizing biases and protecting vulnerable groups, ethical norms play a crucial role in determining the responsible development and deployment of Emotional AI. As developers and policymakers work across disciplines and countries, the formulation of rigorous ethical principles becomes a vital basis for the ethical integration of emotions into artificial intelligence systems.

15.2 The Responsibility of Creators and Manufacturers

The duty of inventors and producers in the field of emotional robotics is a vital element that deserves serious attention. As designers carve the route towards constructing robots capable of subtle facial expressions and emotional reactions, ethical problems come to the forefront. This obligation goes beyond simply technical competence; it covers the larger social implications, possible repercussions, and the influence on human well-being.

Creators are challenged with the ethical problem of defining the limitations and appropriateness of the emotional capacities implanted in robots. The design decisions made by creators bear important consequences for human-robot interaction and impact the dynamics of trust, empathy, and user experience. For instance, a robot developed for companionship in hospital settings must strike a precise balance between offering emotional support and preventing any possible injury stemming from mistaken feelings. Creators must be cognizant of the possible abuse or unexpected repercussions of their works, underlining the necessity for a careful and ethical approach to design.

Manufacturers, on the other hand, play a crucial role in converting conceptual ideas into practical, market-ready items. Their role is not just in assuring the technical operation and durability of emotional robots but also in adhering to ethical norms throughout the production process. This covers factors pertaining to data privacy, security, and the possible social effects of broad usage.

An instructive illustration of the responsibilities of inventors and producers may be found in the creation of emotional support robots for elderly folks. While these robots carry the possibility of reducing loneliness and giving company, developers must traverse the ethical minefield of emotional ties. Manufacturers, in turn, are accountable for establishing protections to preserve the privacy and well-being of the senior users. This entails comprehensive security measures to prevent unwanted access to sensitive data and smart design decisions to avoid possible emotional dependencies that might be destructive.

Furthermore, the task extends to recognizing cultural differences and social expectations. Creators and producers must examine multiple viewpoints to guarantee that emotional robots are culturally sensitive and adaptive to various social circumstances. For instance, a robot developed for emotional support in Japan may need distinct emotions and actions compared to a comparable robot deployed in a Western cultural environment. Failure to notice and accept these cultural variations may lead to misunderstandings or discomfort, stressing the necessity for a globally mindful approach in the invention and production of emotional robots.

The integration of emotional intelligence in robots also raises problems regarding responsibility and culpability. If a robot created to give emotional support mistakenly causes injury or discomfort, defining accountability becomes a tricky matter. Creators and manufacturers must develop explicit standards and processes for resolving such events, including protocols for user education, continuing assistance, and, if required, product recall or discontinuance.

The partnership between artists and manufacturers is important to the ethical development of emotional robots. Regular communication and feedback loops between the design and manufacturing stages are crucial to address developing difficulties and guarantee that the final product fits with ethical standards. Transparent communication channels promote a continuous discourse about possible dangers, ethical issues, and social repercussions, enabling a proactive approach to responsible innovation.

Consideration of the long-term implications of emotional robotics also comes within the attention of designers and manufacturers. As these technologies grow increasingly interwoven into everyday life, the potential for social transformations in human behavior, relationships, and emotional well-being demands continuous monitoring and evaluation. Creators and manufacturers should actively participate in research, cooperation with ethicists, and continual examination of the social effect of their innovations, promoting a culture of responsible innovation.

In conclusion, the obligation of inventors and producers in the field of emotional robotics surpasses technical competency and extends to the ethical, cultural, and social components of human-robot interaction. Through careful analysis of possible implications, adherence to ethical norms, and continual cooperation, artists and manufacturers may traverse the complicated world of emotional robots responsibly. This strategy assures that modern technologies increase human well-being without violating core ethical values.

15.3 Balancing Progress and Ethical Considerations

Balancing progress and ethical issues in the development of emotional robots is a delicate task that demands careful assessment of technology breakthroughs and their possible social implications. As artificial intelligence (AI) and robotics continue to improve, the integration of emotional capacities in machines poses ethical problems that touch upon different elements of human-robot interaction, privacy, and the possible implications of emotional manipulation.

One significant ethical concern focuses around the idea of permission and the bounds of emotional connection. As robots grow increasingly competent at imitating emotions, the issue arises: to what degree should robots be permitted to affect or alter human emotions without express consent? For instance, in the context of emotional support robots meant to cure loneliness, engineers must tread cautiously to ensure that the emotional exchanges are authentic and polite, avoiding any kind of emotional coercion.

Privacy problems often come to the forefront when discussing the creation of emotional robots. The collecting and analysis of personal emotional data for the goal of enhancing AI systems might offer substantial privacy problems. For example, if emotional robots are deployed in public settings to evaluate and react to human emotions, there is potential for the unintended collecting of sensitive emotional data without people' express agreement. Striking a balance between the advantages of emotional AI and maintaining individual privacy becomes a vital ethical concern in the pursuit of advancement.

Furthermore, the possibility for emotional manipulation by robots raises ethical red flags. If robots are built to modify their emotional displays to influence human behavior, there is a potential of exploitation. For instance, in a therapeutic situation when emotional robots are utilized to aid persons with mental health concerns, the ethical barrier becomes blurred. Ensuring that emotional robots behave in the best interest of humans, governed by ethical standards, becomes vital to minimize unforeseen outcomes.

An further ethical problem arises in the possible influence of emotional robots on social interactions. As robots become increasingly integrated into everyday life, there is a concern that over dependence on emotional robots for companionship might lead to a reduction in human-to-human connections. Striking a balance between the advantages of emotional robots and keeping the complexity of human interactions is vital to avert unwanted social repercussions.

In the quest of advancement, developers and policymakers must also address the possible biases encoded in emotional AI systems. If the training data used to educate emotional robots is biased, the computers may accidentally replicate or exacerbate social biases. For instance, if a robot is educated largely on data from a given cultural group, it may fail to effectively comprehend and react to emotions from persons outside that group. Addressing and eliminating these prejudices is a fundamental ethical responsibility to guarantee the fair and equitable deployment of emotional robotics across various populations.

Moreover, the ethical issues of emotional robots extend into the sphere of employment and the possible displacement of human labor. As emotional AI gets more advanced, there is a risk that some tasks traditionally handled by humans, such as emotional support positions, might be outsourced to robots. The ethical concerns here entail managing the difficult balance between technology development and the preservation of job prospects for people. Striking a balance that leverages the advantages of emotional AI without inflicting excessive damage to the labor market is a complicated task that demands serious ethical thinking.

In the sphere of children's relationships with emotional robots, ethical questions take on an extra dimension of complexity. Developers must address the possible long-term influence of emotional interactions with robots on a child's emotional development. For example, if a youngster creates a deep emotional link with a robot companion, what are the ramifications for their understanding of interpersonal relationships and empathy? Ethical rules must be set to guarantee that emotional robots intended for children emphasize the child's well-being and encourage healthy emotional development.

Additionally, the openness of emotional AI systems becomes a major ethical concern. Users dealing with emotional robots should be educated about the fake nature of the emotions expressed by the machines. Failure to give this openness might lead to users forming incorrect views and expectations, perhaps leading in emotional injury when the real nature of the robot's capabilities is disclosed. Striking a balance between providing emotionally engaging encounters and retaining openness is vital to sustain ethical standards in emotional robotics.

In conclusion, the ethical problems surrounding the integration of emotional skills in robots are numerous and need careful navigation. Balancing advances in emotional robotics with ethical standards entails addressing questions of permission, privacy, possible biases, social effect, employment, and the influence on children. Developers, politicians, and ethicists must cooperatively work towards creating clear norms and standards to guarantee that emotional robots contribute positively to society while avoiding possible disadvantages. Only via a careful and ethically led approach will the science of emotional robots attain its full promise in boosting human well-being without violating core ethical norms.

Chapter 16: Reflections and Future Directions

16.1 Lessons Learned from Emotional Robotics

The research of emotional robotics has generated significant insights, giving a rich tapestry of teachings that reach beyond the sphere of technology. One basic lesson learnt is the intricate nature of human emotions and the difficulty involved with recreating them in artificial beings. Emotional robotics has unearthed the complexity of human expression, indicating that emotions are not just isolated responses but intricate interplays of physiological, psychological, and social elements. For instance, recreating the small indicators that transmit emotions, such as micro-expressions or variations in tone, has proved to be a tough undertaking. This underlines the richness and nuance inherent in human emotional experiences, exhibiting the diverse nature that emotional robots must struggle with.

A key lesson derives from the ethical issues interwoven with the development and deployment of emotionally aware robots. As emotional robotics improves, ethical concerns arise, needing a careful balance between innovation and accountability. For instance, the integration of emotional robots in healthcare settings, aimed to give companionship or support, raises problems regarding privacy, permission, and the possibility for emotional manipulation. Addressing these ethical concerns becomes crucial to enable the appropriate and ethical use of emotional robots, highlighting the need for thorough norms and frameworks.

Moreover, emotional robotics has exposed the function of cultural environment in creating and understanding emotional manifestations. The lesson here is that a one-size-fits-all strategy is inappropriate for building emotionally aware robots for various populations. Cultural subtleties considerably affect how emotions are expressed and understood, needing a sophisticated awareness of cultural variety. For instance, a robot meant to communicate with persons from diverse cultural backgrounds must be sensitive to cultural variances in expressing pleasure, grief, or even discomfort. Failure to account for cultural variations may lead to misinterpretations and impair successful human-robot cooperation.

The collaboration potential between emotional robots and sciences such as psychology and neuroscience has emerged as another crucial lesson. By incorporating knowledge from different fields, researchers and engineers may increase the understanding of human emotions, therefore boosting the authenticity of emotional robots. For example, research in neuroscience that dive into the neurological correlates of various emotions might inspire the creation of more precise algorithms for emotional identification in robots. This multidisciplinary approach underlines the connectivity of emotional robotics with wider scientific fields, enabling a more holistic knowledge of both human emotions and the capabilities of emotional robots.

One of the important insights taken from emotional robotics is to the influence of these technologies on society views and attitudes. As emotional robots become increasingly incorporated into all facets of everyday life, public acceptability becomes a vital component. Public impressions are impacted not just by the functioning of emotional robots but also by cultural, ethical, and psychological issues. For instance, the representation of emotional robots in popular media effects how people view and engage these technology. Understanding and addressing public views is vital for the effective integration of emotional robots into society, ensuring that new technologies are accepted rather than regarded with skepticism or dread.

Additionally, emotional robotics has underlined the significance of constant adaptation and learning. The dynamic nature of human emotions demands robots to grow and change their reactions over time. This flexibility is vital for developing long-term partnerships between humans and emotional robots. For example, a robot created to give companionship should be capable of learning about the preferences, emotions, and growing emotional requirements of its human counterpart. This adaptive flexibility is not just a technical prerequisite but also a crucial part of producing emotionally intelligent robots that can actually increase the well-being of humans.

Furthermore, the idea of the uncanny valley has been a significant lesson in emotional robots. Coined by roboticist Masahiro Mori, the uncanny valley depicts the uneasiness individuals may experience when a robot's look and behavior closely match those of a human yet fall short of perfect imitation. This event illustrates the difficult balance necessary in building emotionally aware robots that generate positive rather than disturbing feelings. Striking this balance is vital for building acceptance and beneficial interactions between people and robots, underlining the necessity for intelligent design decisions to traverse the uncanny valley successfully.

The introduction of emotional robots has also underlined the potential for good social influence, especially in sectors such as healthcare and education. Robots developed to aid persons with emotional or cognitive problems have demonstrated promising benefits in boosting well-being. For example, robots offering companionship to the elderly or supporting youngsters with autism in acquiring social skills have showed the ability to make important contributions to human wellbeing. This lecture stresses the revolutionary potential of emotional robotics in solving social issues and boosting the quality of life for various groups.

In conclusion, the path through emotional robotics has been defined by significant insights that transcend beyond the technical complexity of developing emotionally aware robots. These teachings address the delicate nature of human emotions, the ethical issues inherent in emotional robotics, the influence of cultural context, multidisciplinary teamwork, social perceptions, adaptation, the uncanny valley, and the beneficial societal impact of these technologies. Each course adds to a complete knowledge of the difficulties and possibilities posed by emotional robotics, directing the continued development of these technologies toward responsible, culturally sensitive, and morally sound applications in different parts of human existence.

16.2 The Continuous Evolution of Emotional AI

The continual growth of Emotional AI represents a dynamic frontier in the area of artificial intelligence, blending breakthroughs in technology and psychology to infuse computers with the ability to perceive, interpret, and react to human emotions. This evolution is driven by a multiplicity of variables, including the development of advanced algorithms, the inclusion of deep learning methods, and the increased focus on human-robot interaction.

One important feature of Emotional AI's progress is in the refining of emotion detection algorithms. Early efforts frequently relied on simple rule-based systems that failed to capture the subtleties of human emotional responses. However, recent breakthroughs in machine learning, especially in the field of deep neural networks, have substantially enhanced the accuracy and subtlety of emotion perception. For instance, face expression analysis using convolutional neural networks allows computers to recognize subtle emotional signals, opening the path for more genuine and nuanced interactions between humans and emotional AI entities.

Moreover, the continuing progress of Emotional AI goes beyond simply recognition to embrace the production of emotional reactions. AI systems are growing proficient at modeling emotional expressions that go beyond established sets, enabling for more flexible and contextually relevant interactions. For example, chatbots and virtual assistants are combining sentiment analysis to adjust replies depending on the emotional tone of user input, offering a more personalized and compassionate user experience.

The integration of emotional computing, a multidisciplinary discipline that integrates computer science and psychology, plays a vital role in the creation of Emotional AI. By drawing on knowledge from psychology, researchers and engineers may develop AI systems that not only identify emotions but also grasp the underlying context and complexities. This multidisciplinary approach promotes the creation of emotionally intelligent robots capable of reacting properly to a varied variety of emotional states.

In the area of robotics, Emotional AI is affecting the design and functioning of socially supportive robots. These robots are being deployed in numerous fields, such as healthcare and education, to give companionship, support, and help. For instance, robots equipped with Emotional AI may change their behavior depending on the emotional requirements of humans, establishing a feeling of connection and rapport. This progression shows potential for tackling social concerns, like as loneliness among the elderly, by creating emotionally responding robotic companions.

Ethical issues are crucial to the continued progress of Emotional AI. As robots grow increasingly competent at recognizing and reacting to human emotions, ethical frameworks must be built to guide the appropriate development and deployment of these technologies. Issues such as privacy, consent, and the possible manipulation of emotions by AI systems demand thorough investigation and proactive actions to guarantee that Emotional AI serves society without sacrificing core values.

The entertainment sector also acts as a crucible for the advancement of Emotional AI, with applications ranging from virtual characters in video games to interactive storytelling. Emotional AI allows these virtual beings to express a wide variety of emotions, boosting the entire immersive experience for users. For instance, characters in video games may now dynamically react to players' actions, producing a more interesting and emotionally evocative story.

Furthermore, the continuing progress of Emotional AI has consequences for human-robot cooperation in the workplace. As robots and AI systems become fundamental aspects of numerous businesses, their capacity to comprehend and respond to human emotions becomes vital for efficient cooperation. Emotional AI may boost communication between people and robots, providing a collaborative environment where both entities can operate effortlessly together.

The creation of Emotional AI is not without hurdles. The idea of the "uncanny valley," where a robot or virtual entity's near resemblance to human feeling generates discomfort, remains a key challenge. Striking the correct balance between authenticity and avoiding the uncanny valley phenomena is vital for general adoption and integration of Emotional AI into society.

In conclusion, the continuing progress of Emotional AI is a fascinating junction of technology and human psychology. From better emotion identification algorithms to the development of adaptive emotional reactions, this progress is altering human-robot interactions, impacting numerous sectors, and generating critical ethical questions. As Emotional AI continues to grow, its influence on society will extend beyond the world of technology, affecting the way people engage with computers and impacting the ethical landscape of artificial intelligence.

16.3 Envisioning a Harmonious Future Between Humans and Emotional Robots

Envisioning a peaceful future between humans and emotional robots is a multidimensional investigation that digs into the possible cohabitation of artificial intelligence and human emotions. As technology improves, the integration of emotional robots into different facets of our life offers both thrilling opportunities and ethical issues. The vision of a happy future entails not only the smooth integration of robots into society but also the deliberate navigation of problems to guarantee that the connection between humans and emotional robots is balanced, ethical, and favorable to the well-being of both parties.

One crucial component of visualizing a peaceful future rests in recognizing the possible uses of emotional robotics in everyday life. For instance, in healthcare, emotional robots might function as companions for the elderly, offering emotional support and aid with everyday duties. These robots might detect changes in mood and behavior, alerting healthcare professionals or family members to possible concerns. By developing emotional relationships, robots might ease loneliness and contribute to the general well-being of humans.

In education, emotional robots have the potential to improve the learning experience for pupils. Adaptive learning algorithms might employ emotional signals to customize instructional material to individual requirements, offering a more personalized and effective learning environment. Additionally, robots might aid pupils with specific needs, giving emotional support and improving social relationships. The harmonious integration of emotional robots in education might lead to more inclusive and supportive learning environments.

However, the route to a peaceful future is not without hurdles. Ethical issues loom big, necessitating considerable thinking and control. Ensuring the privacy and security of persons engaging with emotional robots is crucial. Striking the correct balance between gathering data for enhancing robot operation and maintaining user privacy is an issue that must be addressed to develop trust and adoption.

Furthermore, the possible influence on the employment market demands careful study. As emotional robots grow more sophisticated, there is the risk of employment displacement, especially in professions needing emotional support or companionship. Ethical frameworks need to be devised to guide the proper deployment of emotional robots, ensuring that their incorporation into society benefits human lives rather than causing damage.

Cultural variances in the acceptability of emotional robots also play a key role in visualizing a peaceful future. Different civilizations may have diverse views regarding the integration of robots into everyday life, driven by cultural conventions, beliefs, and values. Understanding and appreciating these cultural variations is vital to secure the universal adoption of emotional robots and prevent future social issues.

The idea of harmony goes beyond mechanical functioning to the emotional and psychological well-being of humans in the presence of robots. Human-robot contact should be developed to enhance human emotions and connections rather than replacing or reducing them. Emotional robots should be trained to identify and react properly to human emotions, establishing a feeling of connection and understanding.

A major concern in visualizing a peaceful future is the formulation of clear rules for the appropriate development and usage of emotional robots. Collaborative efforts between technologists, ethicists, legislators, and the general public are crucial. Open conversations and inclusive debates may help design policies that address concerns while boosting innovation. For example, organizations like the Institute of Electrical and Electronics Engineers (IEEE) have produced norms and standards for ethical issues in robotics and artificial intelligence, giving a platform for responsible activities.

The representation of emotional robots in popular culture also impacts public perception and acceptance. Media, including movies, novels, and television programs, routinely portray settings where robots either improve or imperil human life. A determined attempt to offer fair and honest images may aid to establishing positive attitudes and minimizing unfounded anxieties.

In conclusion, picturing a happy future between humans and emotional robots involves a comprehensive approach that encompasses technology breakthroughs, ethical issues, cultural diversity, and social implications. The potential advantages of emotional robots in healthcare, education, and companionship are enormous, but the proper deployment of this technology is vital to prevent unforeseen effects. By promoting open debate, developing ethical frameworks, and recognizing cultural diversity, society may move towards a future where emotional robots improve the human experience without sacrificing our values, privacy, or well-being. The goal of harmony entails not only the production of emotionally aware robots but also the nurturing of a social framework that assures their integration corresponds with human values and ambitions.

Refrences

1. Ishiguro, H., & Dautenhahn, K. (2007). Humanoid Robots: A New Kind of Tool. Springer.
2. Picard, R. W. (1997). Affective Computing. MIT Press.
3. Ekman, P. (1992). An Argument for Basic Emotions. Cognition & Emotion, 6(3-4), 169-200.
4. Bartneck, C., Kanda, T., Ishiguro, H., & Hagita, N. (2007). Is the Uncanny Valley an Uncanny Cliff? Robot and Human Interaction, 2007. RO-MAN 2007. The 16th IEEE International Symposium on.
5. Breazeal, C. (2002). Designing Sociable Robots. MIT Press.
6. Fong, T., Nourbakhsh, I., & Dautenhahn, K. (2003). A Survey of Socially Interactive Robots. Robotics and Autonomous Systems, 42(3-4), 143-166.
7. Turkle, S. (2011). Alone Together: Why We Expect More from Technology and Less from Each Other. Basic Books.
8. Billard, A., & Kragic, D. (Eds.). (2018). Trends in Robot Learning. Springer.
9. Brooks, R. A. (1991). Intelligence Without Representation. Artificial Intelligence, 47(1-3), 139-159.
10. Loth, S., & Wrede, B. (Eds.). (2016). Towards a Science of Emotion in Human-Robot Interaction. Springer.
11. Yanco, H. A., & Drury, J. L. (2004). Classifying Human-Robot Interaction: An Updated Taxonomy.

Proceedings of the IEEE/RSJ International Conference on Intelligent Robots and Systems (IROS).

12. Picard, R. W., Vyzas, E., & Healey, J. (2001). Toward Machine Emotional Intelligence: Analysis of Affective Physiological State. IEEE Transactions on Pattern Analysis and Machine Intelligence, 23(10), 1175-1191.

13. Maes, P. (1995). Artificial Life Meets Entertainment: Lifelike Autonomous Agents. Communications of the ACM, 38(11), 108-114.

14. Brooks, R. A. (2002). Flesh and Machines: How Robots Will Change Us. Vintage.

15. Turkle, S. (2005). The Second Self: Computers and the Human Spirit. MIT Press.

16. Bartneck, C., Kulic, D., Croft, E., & Zoghbi, S. (2009). Measurement Instruments for the Anthropomorphism, Animacy, Likeability, Perceived Intelligence, and Perceived Safety of Robots. International Journal of Social Robotics, 1(1), 71-81.

17. Gazzaniga, M. S. (2005). The Ethical Brain: The Science of Our Moral Dilemmas. Harper Perennial.

18. Dautenhahn, K. (2007). Socially Intelligent Robots: Dimensions of Human-Robot Interaction. Philosophical Transactions of the Royal Society B: Biological Sciences, 362(1480), 679-704.

19. Picard, R. W. (2000). Affective Computing: Challenges. International Journal of Human-Computer Studies, 52(1), 89-107.

20. HRI Pioneers. (n.d.). Retrieved from https://humanrobotinteraction.org/community/pioneers/

21. Kismet. (n.d.). MIT Media Lab. Retrieved from http://www.ai.mit.edu/projects/sociable/index.html

22. Knight, W. (2018). Meet the First-Ever Robot Citizen – a Humanoid Named Sophia. MIT Technology Review. Retrieved from https://www.technologyreview.com/s/609232/meet-the-first-ever-robot-citizen-a-humanoid-named-sophia/

23. Billard, A., & Hayes, G. (Eds.). (2019). Social Robotics. Springer.

24. Dautenhahn, K., & Werry, I. (2004). Towards Interactive Robots in Autism Therapy: Background, Motivation and Challenges. Pragmatics & Cognition, 12(1), 1-35.

25. Turing, A. M. (1950). Computing Machinery and Intelligence. Mind, 59(236), 433-460.

26. Wood, R. (2007). In Search of the Uncanny Valley. Proceedings of the 16th IEEE International Symposium on Robot and Human Interactive Communication (RO-MAN).

27. Minsky, M. (1980). K-lines: A Theory of Memory. Cognitive Science, 4(2), 117-133.

28. Suzuki, K., & Kanda, T. (2016). A Robot in My Shoes: Cultural Differences in Cognitive Perspective Taking with a Robot. Proceedings of the 2016 CHI

Conference on Human Factors in Computing Systems.

29. Balkenius, C., Morén, J., & Johansson, B. (2009). A Model of How Emotional Responses to a Robot Depend on Its Lifelikeness. Proceedings of the 2009 IEEE/RSJ International Conference on Intelligent Robots and Systems.

30. Hesslow, G. (2002). Conscious Thought as Simulation of Behaviour and Perception. Trends in Cognitive Sciences, 6(6), 242-247.

31. Hofstadter, D. R. (2007). I Am a Strange Loop. Basic Books.

32. Dautenhahn, K., & Werry, I. (2009). Methodology and Themes of Human-Robot Interaction: A Growing Research Field. International Journal of Advanced Robotic Systems, 6(1), 39-47.

33. Złotowski, J. A., Proudfoot, D., Yogeeswaran, K., & Bartneck, C. (2015). Anthropomorphism: Opportunities and Challenges in Human-Robot Interaction. International Journal of Social Robotics, 7(3), 347-360.

34. Goleman, D. (1995). Emotional Intelligence: Why It Can Matter More Than IQ. Bantam Books.

35. Aamodt, S., & Wang, P. (2008). Modeling Insider Threats Using Social Network Analysis. Proceedings of the 2008 IEEE/IFIP International Conference on Dependable Systems and Networks.

36. Hayes, A. F. (2018). Introduction to Mediation, Moderation, and Conditional Process Analysis: A Regression-Based Approach. Guilford Press.